2009
23/10

D1799756

Design and Manufacture
for Sustainable Development

Acknowledgements

Valuable support, encouragement and professional advice were received from numerous bodies and individuals, especially from the following sponsors:

CIRP
The Institution of Electrical Engineers
The Institution of Engineering Designers
The Royal Academy of Engineering

Organizing Committee

Chairman
Bernard Hon

Vice-Chairman
Richard Dodds

Members

Tim Bullough
Andy Jones
George Moore

Fiona Nightingale
Bob Pond
Chris Wells

International Scientific Committee

L Alting	Technical University of Denmark, Denmark
E J A Armarego	University of Melbourne, Australia
R Bueno	Fundacion Tekniker, Spain
G Chryssolouris	University of Patras, Greece
J Clarkson	University of Cambridge, UK
L Cser	University of Miskolc, Hungary
H A El Maraghy	University of Windsor, Canada
F Jovane	Politecnico di Milano, Italy
H J J Kals	University of Twente, Netherlands
F L Krause	Fraunhofer Institut fur Produktionsanlagen and Konstruktionstechnik, Germany
G N Levy	University of Applied Sciences St Gallen, Switzerland
A Y C Nee	National University of Singapore, Singapore
J Peklenik	University of Ljubljana, Slovenia
G Sohlenius	The Royal Institute of Technology, Sweden
T Suga	University of Tokyo, Japan
N P Suh	Massachusetts Institute of Technology, USA
T Tang	Hong Kong Productivity Council, Hong Kong
S Tichkiewitch	ENSGI, France
H van Brussel	Katholieke Universiteit Leuven, Belgium
D Y Yang	Korea Advanced Institute of Science and Technology, Korea
J Y Zhu	Nanjing University of Aeronautics and Astronautics, China

Design and Manufacture for Sustainable Development

27–28th June 2002 at
the University of Liverpool , UK

Edited by
Bernard Hon

Organized by

THE UNIVERSITY
of LIVERPOOL

Sponsored by

Professional Engineering Publishing Limited,
Bury St Edmunds and London, UK

First Published 2002

© 2002 with Professional Engineering Publishing Limited, publishers to the Institution of Mechanical Engineers, unless otherwise stated.

ISBN 1 86058 396 2

A CIP catalogue record for this book is available from the British Library.

Printed by The Cromwell Press, Trowbridge, Wiltshire, UK

About the Editor

Bernard Hon is the first Professor of Manufacturing Systems at Liverpool University. He is the Director of Product Innovation and Development Centre, and Merseyside TCS Centre for promoting and supporting product and process innovation and knowledge transfer to SMEs. He is also the Programme Director for two MSc courses in Advanced Manufacturing Technology and Product Design. Professor Hon's research interest is focused on new manufacturing technology, manufacturing systems design, and more recently, design and manufacture for sustainable development. Professor Hon has served on numerous national and international bodies such as IEE, Foresight, advisers to overseas universities, and government agencies. He is an active member of CIRP and a Fellow of IEE.

Preface

In a recent meeting in April, over 30 manufacturing professors and academics gathered together to brainstorm topics of fundamental impacts to manufacturing in the next twenty years. The result was unanimous: it was sustainability.

To achieve sustainability, we need to understand and deploy the process of sustainable development. Since the, now famous, Brundtland definition published in 1987, our understanding of sustainable development has progressed further. For instance, we now know that there are three pillars of sustainable development, i.e., technical, economic, and social. Unlike other engineering disciplines such as aerodynamics, sustainable development requires a concurrent view of all three factors for achieving an acceptable solution.

The idea of achieving sustainable development is evolving. While the attempt to use the phrase 'to get more from less' to represent the whole essence of sustainable development is inviting, it does not serve real practical purposes. For manufacturing engineers who are responsible for all man-made products, we need deeper understanding and better tools to ensure that our future generations will not be deprived of satisfying their own needs.

This book is the result of the First International Conference on Design and Manufacture for Sustainable Development held at Liverpool University from 27–28 June. The principal purposes of this conference are to:

- stimulate technical and scientific discussions on sustainable development for the manufacturing industry;
- provide a focus for the exchange of the latest ideas and developments on sustainable product life cycle from design to disposal;
- evaluate the total impact of sustainable development principles to product design and manufacture.

This is a small step towards our effort in searching for design and manufacturing solutions which do not exist today. Through international exchange, discussion, and debate, we shall make further advances towards our journey to sustainability.

Bernard Hon
University of Liverpool, UK

Contents

Keynote Papers

Concept, context, and co-operation for sustainable technology

J C van WEENEN
Sustainable Technology, Alkmaar University of Professional Education, and Sustainable Product Development, IBED – University of Amsterdam, The Netherlands

Abstract

Sustainable technology is part of the broader concept of sustainable development. This paper presents a position for sustainable technology in the scientific and societal debate. It emphasises the importance of definitions, concepts, principles and systems on the one hand and of their contextual practical implementation on the other hand. Essential to the success of sustainable technology are integration, participation and co-operation.

Taking contextual initiatives for sustainability while at the same time connecting to and co-operating with projects in developing countries, seems an attractive and new way forward. This will be demonstrated by three sustainable technology examples from different parts of the world. Finally an observation will be presented concerning international trends and future development.

1 CONCEPT, CONTEXT AND COOPERATION

1.1 Concept

Sustainable Development is an integral concept for achieving quality of life. Life is about "L" for Limits, "I" for interdependence, "F" for fundamentals and "E" for equity. This set of issues reflects the importance of dealing with material concerns, acknowledging the relationship between humanity and nature, being committed to addressing fundamental causes, and considering ethical values (1). Humanity therefore must formulate totally new principles for fundamentally different systems of production and consumption, including:

- Substantially lower levels of material and energy intensity.
- Much less throughput.
- Much more use of the sun and renewable materials for inherently sustainable processes, products and services.

At the basis of these principles is Sustainable Product Development (SPD) which is defined as resource, context and future oriented product development, aimed at the fulfilment of elementary needs, better quality of life, equity and environmental harmony (2).

According to UNDP, consumption must be shared, guaranteeing basic needs for all, it must be strengthened by building human capacities, and it must be sustainable, without mortgaging the choices of future generations. UNDP recommends the development and application of technology and methods that are environmentally sustainable for both poor and affluent consumers. Sustainable growth of consumption and production depends on major advances in cleaner, material-saving, resource-saving and low-cost technologies that meet the requirements of the poor (3).

1.2 Context

Technology is the whole domain of knowledge, concepts, theories, techniques and methods, procedures, organisations and processes, which serves to prevent, resolve or reduce problems. Sustainable Technology is here defined as technology which has the objective to develop, design, apply, evaluate and improve solutions that are aimed at meeting elementary needs on the basis of natural resources, in their local and regional context. It is characterised by a very selective, minimal and whole use of preferably renewable resources, with social-cultural integration, in harmony with nature, while giving account of global relations, a fair distribution and striving for 'co-development' with developing countries and focussing on the future. At the basis and central to sustainable technology are definitions, concepts, principles and systems on the one hand and of their contextual practical implementation on the other hand.

1.3 Co-operation

Essential to the success of sustainable technology are integration, participation and co-operation. Therefore, sustainable technology innovation is inextricably linked to societal process innovation. Regionally this can be realised by contact networking of proponents and pioneers, and internationally by virtual networking which is a relatively new and very promising means. Through integration of local and regional considerations with universal principles and global ideas, an enormous potential of new and sustainable options can be developed. Taking contextual initiatives for sustainability while at the same time connecting to and co-operating with projects in developing countries, seems an attractive and new way forward. This will be demonstrated by three sustainable technology examples from different parts of the world.

2 SUSTAINABLE TECHNOLOGY EXAMPLES

2.1 Zero waste date palm material system, Egypt

The date palm is one of the most important plants in Egypt and in the whole Arab World. Its fruit is dates, high energy food for both people and livestock. In addition to dates, the palm tree provides several secondary products such as leaves, coir, stems, kernel and trunk. Palm trees must be regularly pruned, removing old, dead or broken leaves. These then are used as a by-product for making crates, ropes and baskets, as fuel for heating and cooking and even for house construction. The purpose of the pruning is to clean out the tree, thus allowing new leaves to photosynthesise and grow. It also reduces rodent and insect infestation and facilitates harvesting and the use of leaves as by-product material. The fibre of the leaf base can be used as well. The pruning improves the quality of the crop by reducing shade and bruising of fruits. When the palm is pruned, the offshoot and the sucker on the crown of the tree are also removed.

Date, the primary product of the palm, is rich in protein, vitamins, and mineral salts. In countries such as Egypt it represents an essential element in the diet of the cultivator and of animals, for which the low-grade date with kernel can serve as food. All secondary products of the palm result from annual pruning and present an essential use potential for the cultivator. Thus the palm grows without generating any waste.

The date palm's midribs or grown palms are used as roof material, after having been woven as a mat, using coir ropes. The palm midrib is also used to make crates for the transportation of vegetables and fruits, furniture items, manual fans, garden gates and coops for chicken and rabbits. Midribs of the young palms are used for fencing gardens, for floats for fishing nets, or for fuel in rural ovens. Afterwards the ashes are used in mortar.

Hamed El-Mously, Director of the Centre for Development of Small-Scale Industries in Cairo, Egypt, considers the various uses of the secondary products of the palm tree to be eloquent examples of local innovations. Since 1990 he conducts research on the properties and use potential of the midribs of the date palm leaves (4). He rightly finds that the date palm in Egypt is a good example of integrated sustainable use of renewable material resources. He is of opinion that as a Third World researcher his task is to direct imagination and thinking to the formulation of a new vision on the use of local material resources. This he calls "the rediscovery of our natural resources".

Traditionally, Date Palm Leaf Midrib (DPLM) was mainly used for ceilings and fencing in rural areas, as well as for crates used to transport vegetables. According to El-Mously, with the changes in the way of life in Egypt and in the region, these uses diminished, which led to the neglect of pruning and a decrease in the economic feasibility of palm cultivation. If pruning did occur, the resulting leaf waste became a burden to the environment. Against that background, development activities have been directed to application of DPLM in final product lines as a substitute for imported wood.

One product line is Arabesque (Mashrabiah) hand craft products, which used the midribs as a substitute for beech wood. A drastic increase in the price of imported beech wood has led to a diminishing demand for Arabesque handicrafts. Therefore, it was thought that replacement of beech by cheap, locally available palm leaf midribs might lead to a revival of Arabesque handicrafts, especially in rural areas.

The Centre launched a project in 1995 for the development of small-scale industries in the Dakhla oases in the New Valley governate, to disseminate Arabesque handicrafts based on palm midribs as a raw material. A training centre was established to train workers, who receive their palm lathes on a loan basis for production at home. The project, El-Mously claims, was a great success as it turned the poor, especially women, into autonomous producers and entrepreneurs, while changing the idea of using tree pruning products as a substitute for imported wood, into reality. Thus the project reveals the great potential for a new culture of sustainable use of renewable material resources in rural and desert communities in the whole region. El-Mously's work started from the basis of the local community as a socio-cultural-ecological system with an identity of its own. Development, or indigenous development, he argues, means to stimulate the people's own desires for improvement in all ways, and their potential energy and creativity to create conditions for maintaining these transformations.

El-Mously evaluates that one of the most interesting developments was that the beneficiaries - especially women - began to use other materials derived from the pruning of other fruit or wood trees in their vicinity (olive trees, lemon, casuarina, etc.). They dried the raw material in the open courts of their houses and used it for the production of a wide variety of products. It underlined that through the introduction of the use of DPLM the message got across that new uses for surrounding resources can always be found and that nothing in nature is without value, although technical support by designers would remain a requirement.

On the basis of this experience, El-Mously defines appropriate technology as those technical inputs (tools, machines, techniques, training courses, knowledge, etc.) whether endogenously initiated, or introduced from elsewhere, that could be assimilated by the recipient community so that this may lead to building or strengthening its technological capabilities. With the development of the local community, its capability for assimilation would grow. Thus, according to El-Mously, the perception of the local community as a living and growing entity, gives a sense of dynamics to the definition of appropriate technology.

He calculated that during its life, a date palm represents a crop of lignocellulosic material more than ten times that of the spruce, in addition to the dates, of course. This Factor 10 alternative, he suggests, opens the way to a new concept of afforestation more appropriate to the Middle East: afforestation to obtain food and lignocellulosic materials that may serve as wood substitutes and other industrial uses. Besides and perhaps more important, the date palm points to a new ethos from an environmental perspective: one can obtain "wood" not by felling trees, put by serving (pruning) palms. Thus the palm can play an important role in a sustainable future. He therefore refers to the date palm as 'The Princess of a Sustainable Future'.

2.2 Zero emission cycle rickshaw system, India

In the city of Agra in India a new model of cycle rickshaw was introduced in 2000. It was built in co-operation with the Institute for Transportation and Development Policy (ITDP), a non-governmental organisation in New York which promotes environmentally sustainable projects world-wide, and the Asian Institute of Transport Development (AITD) in Delhi. The user-friendly rickshaw is one of the six models they developed. The project began in 1995 and was aided by the United States Agency for International Development (USAID) with design and technical support from the Indian Institute of Technology (IIT) in Delhi.

Agra in Uttar Pradesh, is the city of the world famous Taj Mahal. One of the aims of the cycle rickshaw improvement project was to save the Taj Mahal that has been affected by air pollution mainly from motor vehicles and internal combustion power generators. The funding for the project was part of an environmental effort to save the Taj Mahal, a site of world cultural heritage. Although motorised vehicles are banned within a 4 km radius of the monument, air pollution still affects this building.

According to Hook, executive director of ITDP, this institute was concerned about the environmental ramifications of the trend in India of rapid increases in motor vehicle use (5). As incomes rise, he observes, individuals switch from using the non-polluting, energy-saving bicycles and non-motorised cycle rickshaw to highly-polluting two-stroke engine motorcycles and the three wheeled Bajaj taxi, resulting in a dramatic increase in transport air emissions. As a consequence, residents of Indian cities face severe public health risks as a result of exposure to levels of air pollution well above World Health Organisation standards.

Motor vehicle emissions are also a source of CO_2 emissions, which are a major cause of global warming. Hook points out that although currently India is not a major contributor to greenhouse gas emissions, if the country motorises rapidly using available technology, their contribution to global warming will grow sharply. Against this background, damage to monuments is just another adverse effect of local air pollution associated with motorised traffic. ITDP was therefore looking for ways of encouraging the use of human powered taxis, cycling, and walking in population centres where the air quality benefits are the most important, and where distances are short enough to make this mode economically and commercially viable. In order to do so, however, as Hook underlines, many criticisms of cycle rickshaws both legitimate and unfair, would need to be overcome.

The cycle rickshaw is an important source of employment for many of India's lower income people. In Agra the cycle rickshaws provide employment to roughly 10% of the adult population, and a critical course of low-cost basic mobility for India's lower and middle classes. Hook writes that even wealthier Indians tend to rely on the cycle rickshaw for bringing their children to school, paying a monthly fee for the service. At the same time, there is a stigma attached to being a rickshaw puller a 'wallah', what is considered a lower class job. Agra is also an important city of regional manufacture of cycle rickshaws and is known for producing a high-quality cycle rickshaw. Agra produces between 30.000 and 50.000 cycle rickshaws per year, the majority of which are exported throughout the region, within a roughly 80 km radius of the city. Hook observes that despite the environmental benefits of the cycle rickshaw, and the economic importance of the industry particularly to low income families, public attitudes towards the cycle rickshaw in India are negative, and their use is being phased out in many Indian cities and elsewhere in Asia.

Rickshaw pullers have many reasons to opt for the new cycle rickshaws (6). The new model reduces strain and fatigue and allows them to work for a longer period of time. According to designer Gadipalli Shyam, they weigh only 55 kg as compared to the traditional ones that weigh 90 kg. Another advantage is the addition of two new gears, which reduce the pedalling effort by 17 %. These new gear systems manufactured by Spark Engineering of Ghaziabad in Uttar Pradesh, cost Rs 400 and can be fitted to traditional rickshaws as well. According to Matteo Martignoni, vice-president of ITDP compared to the traditional ones that usually last for two to three years, the new ones are expected to last for nine to 10 years. The new models are not very expensive. They cost about 10-15 per cent more than the conventional ones. The convenience to the passengers has also been improved. The footboards of the new rickshaws are placed comfortably low and the seats have been widened. The canopy is very well placed and provides good protection against the sun. The luggage space is another reason why passengers are opting for the new rickshaws. The chassis are lower and longer. This adds stability and reduces the centre of gravity. The rear wheels are also equipped with chassis to decrease friction and wear.

The India Cycle Rickshaw Modernisation Project ended on October 31st, 2000. In February 2001 there were roughly 700 modern cycle rickshaws on the streets of Agra, Vrindavan, Mathura and Delhi. The redesigned vehicles became popular with drivers particularly around the Taj Mahal in Agra and the Krisna Temple in Vrindavan, not just because they weighed much less than the traditional ones, but because they showed an air of modernity that reinvigorated people's willingness to travel by pedal power. It demonstrates, according to Martignony, that "modernisation doesn't mean motorisation".

As there had been no successful implementation of innovative designs in cycle rickshaws in the last two decades, the challenge to make major improvements was quite hopeful. However, as Shyam commented in an interview: "It is easy to make things complicated, but very difficult to make things simple." A do-it-yourself version of the project is available on the Internet to preach the benefits of cycle rickshaws in areas where they are not used and to introduce innovations where the traditional models are prevailing.

The newly designed rickshaws were expected to be also used in the US and in European countries. The commercialisation of the modern rickshaw system seems to have been successful as a similar project began in Indonesia in the spring of 2001.

2.3 Zero energy sustainable school system, the Netherlands
One of the basic requirements for the realisation of the process of sustainable development, is sustainable education. All around the world initiatives have been taken to integrate sustainable development into the activities of existing educational institutions such as universities and professional schools. Also the integration in secondary and in elementary education is emerging.

With respect to sustainable building of educational facilities, the focus has mainly been on energy saving, although some educational buildings have been realised on the basis of a broader scope. Totally new approaches can be found as well, in which the development of sustainable education is integrated into the building process of a new school or a new educational facility.

In November 1997 the municipality of Castricum, the Netherlands, decided to build a new school building for the public elementary school De Sokkerwei because of the deplorable state of its two old school buildings. In March 1998 ICS consultants were commissioned to write a discussion note for and in participation with the schoolteachers, management and board on the characteristics, vision, mission and requirements of the school.

At that time it had become clear that for the realisation of the school building other building partners would be required. Contacts were made with the local organisation for after school care (NSO), and with local housing associations. In February 1998 the School Board asked Hans van Weenen as a parent and former member, to investigate the possibilities of sustainable building of a new school. In February 1999 he completed a discussion report on the subject.

He defined a sustainable elementary school as a sustainably designed, built and used school building in which sustainable education is being developed and provided. The building and its surroundings serve as an educational tool for illustration, demonstration, exploration, experimentation and discovery.

Project developer Lithos Bouw B.V. was chosen to build the whole plan consisting of a school, school apartments, an apartment building and free standing houses. An agreement indicated that an additional budget of 5-10% of the building costs of the school, resulting from the exploitation of the area, should be reserved for the sustainable building of the school. The municipal council finally approved the plan on October 4th 1999, unanimously supporting the sustainable building of the plan, and especially of the school. Then the architect BBHD was selected and the design process began in November 1999 (7).

The consultancy IDEA was appointed as sustainable building advisor. The Netherlands Energy Research Foundation (ECN) became involved to do research on a zero-energy concept for the school, with support of NOVEM. The Dutch Governmental Building Agency (RGD) decided to give the project the status of National Demonstration Project on Sustainable Decision-Making. It provides support for maintaining and if possible upgrading the formulated ambition level. The Province of North-Holland nominated the project for the Dutch National Future Prize 2000. IDEA and EDEN developed a web-site on sustainable educational buildings for sustainable education with information, links and cases from around the world.

TNO-Centre for Timber Research investigated a timber frame building system for the school as well as the application of other renewable resources-based materials. They also addressed the reuse of constructions and materials from the two existing school buildings. Furthermore, international workshops were organised on passive solar schools and on sustainable water management in support of the development of the new Sokkerwei building and school grounds.

One of the objectives of the project was to realise a zero-energy school. 'Zero-energy' in this case means 'CO_2 neutral'. Central to the concept involved is to achieve the largest possible saving in the use of fossil fuels and to compensate the consequences of the remaining fossil energy consumption by using sustainable energy. Problems concerning tight financial means for school building and in particular for the implementation of sustainable building measures were resolved by a joint will to achieve the objective. Architect office BBHD and the Dutch Energy Research Foundation ECN designed the building and the technical installation (8).

The school consists of nine classrooms, a common space and a section for after school care. The volume of the building is approximately 5.200 m^3. The insulation value of the front, roof and floor is subsequently 3.5 - 4.0, and 3.0 (m^2K/W) which is substantially better than usual. The HR++ glass in the front and the roof has a U-value of 1.2 (W/m^2K). Part of the outer shell of the school doesn't loose heat because of the apartment building to which it is connected. The calculated energy usage is 4.000 m^3/y of gas as compared to 9.300 m^3 for a standard school and 14.600 kWh/y of electricity with 26.000 for a standard school. This low energy usage is the result of a compact, well insulated, round building design, good day lighting with daylight dependent electronically controlled energy saving lighting, a balanced ventilation system met HR ventilators and HR heath recovery. These integrated systems are controlled by a building management system.

The school will use solar cells which will linked to the electricity net, in order to attain a CO_2-free production of electricity that is associated with the building. The CO_2 emission which results from the use of gas will be compensated by using windmill electricity supplied by the regional wind energy cooperation Kennemerwind. Due to unexpected substantial rising of building costs in the spring of 2001, the whole project entered a critical stage. The financial hurdle of building the school was overcome but realisation of the zero-energy concept could not be completely secured.

A multiyear process of integration and development of sustainable elementary education for De Sokkerwei has begun. Students of Alkmaar University of Professional Education developed a zero-energy computer game for the children of the school to clarify the concept to them.

Currently an evaluation report on the project from January 1997 - January 2002 is being drafted for RGD. The first sustainable elementary school of the Netherlands will be completed in July 2002. Time will tell whether or not all of the original objectives have actually been achieved.

3 INTERNATIONAL TRENDS AND FUTURE DEVELOPMENT

The Internet provides an enormous variety of sources of information on sustainable technology. Regularly searching for new information on that subject, one can obtain an overview of its content and on trends. Thus one can identify and proceed to different levels of sustainable technology definition and practice, as if going up the stairs of the solar temple of sustainable development.

It is striking, however, that many sustainable technology initiatives concern activities previously covered by concepts such as 'environmental improvement', 'environmental management' or 'pollution prevention'. This can be called a first level strategy, which actually is about renaming a rather traditional practice, without really changing anything.

At a second level initiatives can be placed that aim to optimise existing products, processes and systems by integrating sustainable components such as solar energy technology or by changing the process input by using a renewable material resource.

A third level consists of changing from one concept or system to an already available alternative one which is more sustainable by design. This would involve both technological as well as organisational changes.

Finally there is the top level, closest to the sun, and to the overview on a sustainable societal landscape, which is concerned with the development and design of totally new, fundamentally sustainable systems. This is demonstrated by the examples above. Other examples concern sustainable SMEs (9) and renewable material resources systems (10). In such examples a holistic height is attained, in which quality of life is within reach of everyone, the sun can be enjoyed and there are very nice views and perspectives on the future.

REFERENCES

1   **Weenen, H. van** (2000) Towards a vision of a sustainable university. In International Journal of Sustainability in Higher Education, Vol. 1, No. 1, pp. 20-34.

2   **Weenen, J.C. van** (1997) Sustainable Product Development: Opportunities for Developing Countries. In Industry and Environment, UNEP-IE/PAC, Paris, January – June, pp. 14-18.

3   **UNDP** (1998) Human Development Report 1998, Oxford University Press, Oxford.

3   **El-Mously, H.** (1998) The Date Palm: the Princess of a Sustainable Future, In INES Newsletter, No. 23.

4   **Hook, W.** Taj Mahal cycle taxi improvement project: an NGO-private sector partnership. ITDP, New York (http://www.workbike.org/research/tajmahal.html).

6   **Anonymous** (2000) Green drive, In Down to Earth, March 15, pp. 52-53.

7   **Weenen, J.C. van, Dettmers, W.J.M., Overtoom, M.S.J.,** and **Poldermans, H.G.J.M.** (2000) Development of the First Sustainable Elementary School of The Netherlands, In International Conference Sustainable Building 2000, Proceedings, 22-25 October, Maastricht, the Netherlands, pp. 651-653.

8   **Dettmers, W., Sijpheer, N.,** and **Sjoerdsma, E.** (2002) Nul-energie basisschool. Duurzame basisschool 'De Sokkerwei' te Castricum, Nederlandse Duurzame Energie Conferentie 2002, poster, 28 februari 2002.

9   **Weenen, H. van** (1999) Design for Sustainable Development. Practical Examples of SMEs, European Foundation for the Improvement of Living and Working Conditions, Dublin.

10 **Weenen, J.C. van** (2001) Renewable Material Resource Systems for Sustainable SMEs. Sustainable Product Development and Sustainable Consumption in Developing Countries. Research Report, Commissioned by the Ministry for Housing Physical Planning and the Environment, IBED - University of Amsterdam, March 2001.

Methods and elements towards sustainable products – BMW's strategy in design for recycling and the environment

W FRIED, T HAGEN, and **G WÖRLE**
BMW Group Recycling, Munich, Germany

ABSTRACT

The BMW Group sees and acknowledges the three pillars of sustainable development as a fundamental factor in competition. As a company pursuing a policy of sustainable management, we are committed to the ongoing, continuous enhancement of our corporate value in economic, social and ecological terms.

With the areas of Design for Environment (DfE), Design for Recycling (DfR) and production-related environmental production (EMAS, ISO 14000) increasingly growing together to form an integrated product policy, the focus on specific problem areas is being replaced increasingly by a new philosophy of "lifecycle thinking". We also see, however, that external factors are obstructing the ecological optimisation of products to a growing extent.

One example is carbon fibre (CFRP), where we see how the use of a lightweight material able to optimise the overall ecological balance by offering a very substantial improvement of fuel economy during the lifecycle of a car, is ruled out in practice by the European End-of-Life-Vehicle (ELV) Directive requiring 85 per cent recycling of materials.

1. THE PRINCIPLES OF SUSTAINABLE DEVELOPMENT WITHIN THE BMW GROUP

The principles of sustainable development have been part and parcel of the BMW Group for a long time. Proceeding from the Group's success in the market, BMW is consistently enhancing its high social standards. The success of this corporate philosophy is borne out, first, by the fact that in the year 2002 the BMW Group was acknowledged by *manager magazin,* one of Germany's leading business journals, as the most popular employer in Germany. The second factor is the high standard of motivation and performance among BMW Group associates providing the very foundation for success in the market.

Permanent improvement of environmental standards at BMW Group plants and with BMW Group products is seen as a natural asset. A further indicator underlining the success of BMW's sustainable mobility concept is the consistent leadership of the company in the Dow Jones Sustainability Group Index.

Outstanding examples of the BMW Group's responsibility to both society and ecology are the CleanEnergy concept using hydrogen recovered in a regenerative process as the source of energy for the future and the high standards of social and environmental competence at BMW plants applied the world over.

For many years, finally, the principles and tools of Design for Environment (DfE) and Design for Recycling (DfR) have been among the standards acknowledged and applied in the development of new cars and motorcycles.

2. RESPONSIBILITY FOR OUR ENVIRONMENT – THE BMW GROUP'S ENVIRONMENTAL POLICY

"Sustainability is becoming a fundamental factor crucial to economic and social prosperity as well as the interaction of the market and democracy. The BMW Group will maintain its focus on sustainable development as a future-oriented principle in its corporate strategy." (BMW Board Resolution, 21 February 2000)

As a genuine global player, the BMW Group follows the concept of sustainability. An important feature in implementing this vision is that all of BMW's production plants apply the same environmental standards. They are certified to the international DIN ISO 14001 environmental management systems and to the European Eco-Management and Audit Scheme (EMAS). The principle of sustainability is also borne out by BMW products through their concept of sustainable mobility, a concept developed by the BMW Group connecting ecological, economical and social factors with one another.

Sustainable mobility must promote both individual quality of life and economic prosperity. At the same time any impairment of the environment caused by individual mobility must be reduced to a minimum. This makes it essential, first, to sensibly connect the main means of transport with one another, allowing each means of transport to provide and use its individual strength. The second point is that the BMW Group focuses on the ongoing development of conventional drive systems as well as the development of alternative drive concepts for motor vehicles in a process leading to the supply of carbon-free energy. Within the CleanEnergy Project, the BMW Group advocates the cleanest fuel in existence: hydrogen recovered in a regenerating process.

Sheer driving pleasure in the ultimate driving machine - BMW remains committed to this philosophy, which is indeed part of the sustainable mobility concept. A car helping to preserve the environment must maintain current standards in terms of comfort, performance, equipment, use and safety, at the same time meeting the individual demands and requirements of the customer. Otherwise a car would not be successful in the market, since low fuel consumption alone, even in days of high fuel prices, is not an adequate purchasing motive.

The BMW Group therefore seeks in particular to reduce fleet consumption and at the same time develop competitive products offering sustainable mobility. It is not BMW's intention to develop cars devoid of any practical value to the customer, such as the three-litre car or even the one-litre "economobile". Electric cars using fuel cell technology are likewise not able to offer a sensible alternative in the long run due to the weakness of their technical concept. On the other hand the BMW 750hL, a hydrogen car already close to production standard, is able even today to combine environmental protection with a high standard of motoring comfort and superior function. Running on hydrogen, the BMW Group's 750hL emits nothing but steam when the engine is running.

Table 1: BMW's Model of Sustainable Mobility

Intelligent networking of various transport providers	Continuous reduction of average fleet consumption
- making optimum use of each individual transport provider. - reducing traffic volume and making optimum use of transport space.	- in order to preserve finite natural resources - and to minimise exhaust emissions from motor vehicles.
Recycling-optimised product concepts and return/re-use of end-of-life vehicles - in order to preserve resources - and to avoid/ reduce waste.	**Development of alternative drive concepts** - ensuring long-term maintenance of our quality of life and the economic use of individual mobility. - moving step-by-step towards the global use of hydrogen in motor traffic and the vision of a regenerating hydrogen cycle.

Fig. 1: Elements of the BMW's Model of Sustainable Mobility

3. HOLISTIC APPROACH – LIFECYCLE THINKING

Various examples in the past have shown that while political and corporate activities can provide local improvements within the environment, these improvements simply meant a shift of the environmental burden to other sectors of the economy and political life (for example the use of aluminium as a lightweight material in the automobile, shifting the consumption of energy from the use phase to the production phase). Industry and the economy as a whole are therefore focusing increasingly on an overall perspective comprising the entire lifecycle of a motor vehicle.

Fig. 2: Lifecycle thinking right from the start in the design phase

In consideration of the above, the BMW Group ensures that each new model series fulfils the requirements of environmentally-oriented product development by consistently integrating environmental and recycling concepts into the target agreement process from the very beginning. From the start, even in the early development phase, the overriding objective is to minimise the effects of a vehicle on the environment throughout its entire lifecycle.

To reach this objective, the BMW Group uses and intelligently networks the following tools and methods of environmental management [1,2,3]:
  - Design for Environment (DfE),
  - Design for Recycling (DfR)
  - Environmental management systems to ISO 14001 and EMAS (production plants and dealer network)
  - Material flow management
  - Life Cycle Assessment (LCA)

Close cooperation and communication within the chain of players involved, particularly with suppliers, customers and recyclers, is a further important building brick in this process.

The areas mentioned above are increasingly growing together to form a concept of integrated, product-related environmental management. The old philosophy of thinking in specific problem areas being replaced increasingly by a new concept of lifecycle thinking. While, admittedly, a lot of education and enlightenment, as well as detailed work, still remains to be done within the company, it is a fact that external factors are increasingly preventing the development and production of ecologically optimised products, as is described by an example in Chapter 5.

Table 2: Environmental improvements of the automotive lifecycle (eg with the new BMW 7 Series [4])

DEVELOPMENT
- Development of recycling-optimised components
- Clearance of components for the use of secondary materials
- Cooperation and communication with system suppliers on environmental relevant issues
- Construction of environment-optimised components by way of lifecycle assessment
- Reduction of fuel consumption and CO_2 emissions versus the previous model by
- developing consumption-optimised engines
- introducing sophisticated, quality-oriented lightweight technologies
- improving the standard of streamlining
- Reduction of noise emissions
- Use of regenerating raw materials
- Ongoing development of hydrogen drive with a production model already planned

PRODUCTION
- Environmental management system certified to ISO 14001 and EMAS
- Use of water-based base coat and powder clear paintwork
- Implementation of production waste recycling concepts
- Consistent use of quality-proven secondary raw materials

OPERATION, SERVICE
- Fuel consumption under the EU standard just 10.7 or 10.9 ltr/100 km (26.4/25.9 mpg)
- Condition-based service and lifetime oil charge reducing the consumption of operating fluids and parts subject to wear and tear
- Environmentally-friendly disposal of waste materials in service through the BMW workshop waste management system
- Certification of BMW Branches to ISO 14001

RECYCLING
- End-of-life vehicles returned to BMW dealers and certified dismantlers
- Environmentally-friendly recycling of vehicles
- Auditing of dismantlers to improve the quality of recycling
- Reconditioning of components for the exchange scheme ensuring efficient recycling
- Establishment of ecological and economical material cycles
- Production of high-value recycled materials for vehicle production
- Operation of the Recycling and Dismantling Centre (RDC) as a know-how clearing house on all recycling issues

4. DESIGN FOR RECYCLING AND LIFECYCLE ASSESSMENT – TOOLS FOR SUSTAINABLE CAR DESIGN

"It is our objective to promote vehicle design suitable for recycling as well as the use of secondary raw materials, minimising the overall consumption of energy and resources in production and operation and closing complete cycles in the re-use of materials." (BMW Group Environmental Guidelines, item 7).

A modern car must serve many purposes: It must offer superior performance on minimum fuel, adequate space despite compact exterior dimensions, attractive, modern design, low weight, a variety of functions, appropriate options for recycling, a high standard of reliability and all-round safety - and, not least, all this must come at a competitive price. To meet all these requirements, the BMW Group is constantly improving both processes and the workflow in vehicle development.

4.1 Development: processes and methods

Apart from conventional features such as safety, quality and comfort, the overall list of specifications and demands for a new vehicle comprises standards and requirements ensuring a high level of environmental compatibility. This also means Design for Recycling, Life Cycle Assessment, as well as the improvement of exhaust and noise emissions.

New developments are accompanied within the BMW Group by a process of gateway management – a standardised planning system binding for everybody involved in a specific project and extending all the way to individual components provided by suppliers. Disciplined maintenance of the major cornerstones in such a project is essential for punctual processing and management leading up to the start of series production. One of these milestones is the "package freeze", i.e. the point in time at which the space required by all components is determined once and for all.

The design process involving the development of various alternatives, the choice and determination of the final models ends with the "design freeze". In contests serving to choose the most suitable suppliers, the BMW Group is able to plan new technologies and innovations in good time. Preparation of prototypes also within good time serves to provide appropriate test results when they are needed, allowing the responsible engineers to verify the series production process. The improved exchange of information is attributable, not least, to the integration of suppliers at an early point, suppliers even receiving workstations at BMW's Research and Innovation Centre to ensure their proper integration.

4.2 Instruments for sustainable design

Environmental and recycling requirements must fit perfectly into this Product Evolution Process (PEP). Fig 2 therefore presents the gateways in the Product Evolution Process in which environmental and recycling requirements are integrated and taken into account. These requirements are treated in the same way as other product requirements, the due observance of target factors being verified time and again in the ongoing process. Wherever a conflict of interest arises, therefore, a solution is found through a standardised trouble-shooting procedure.

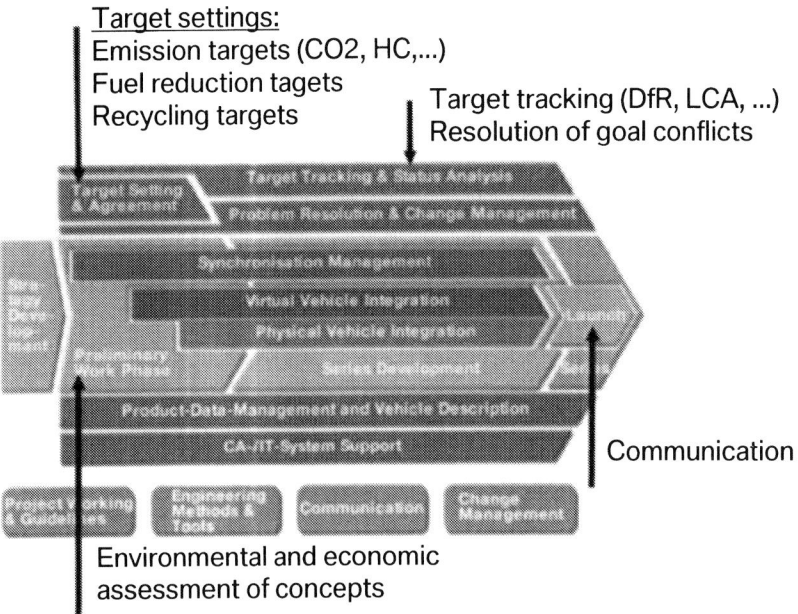

Target settings:
Emission targets (CO2, HC,...)
Fuel reduction tagets
Recycling targets

Target tracking (DfR, LCA, ...)
Resolution of goal conflicts

Communication

Environmental and economic
assessment of concepts

Fig. 3: Environmental and recycling requirements in the Product Evolution Process

4.2.1 Design for Recycling

BMW started its first recycling activities back in 1989, thus becoming one of the first car makers to assume responsibility for the product also beyond the actual period of use. An important step was taken in 1990 with the construction of a pilot dismantling facility in Landshut, approximately 70 km north-east of Munich. Then, proceeding from the know-how obtained at this pilot plant, the decision was taken in 1993 to build a Recycling and Dismantling Centre (RDC) near BMW's Research and Innovation Centre. The reason for this decision was to have a special corporate unit optimising and consistently enhancing the standards of vehicle recycling in terms of both re-use and product development.

Reflecting the BMW Group's holistic product philosophy, this establishes a complete, self-contained cycle of information for all players involved in the recycling process - from development all the way to the re-use of components and materials. This integrated approach taken by the BMW Group regards development, production, use and – at the end of a vehicle's life – recycling as one complete system. Today, the RDC is a unique know-how clearing house covering both Design for Recycling and the efficient use of end-of-life vehicles in ecological terms.

Apart from focusing on ecologically-compatible re-use of end-of-life vehicles, the RDC gives particular significance to the consideration of environmental criteria and recycling standards in the early phase of vehicle design and development. The main ecological development targets are therefore set forth clearly in the initial phase of defining every new model, in addition to technical and economic targets.

These objectives must then be defined even more specifically in the process of product development. To this end, the Recycling and Dismantling Centre assesses previous and competition models in dismantling analyses, determining specific targets for various components.

Fig. 4: Dismantling analysis of the BMW X5 (components relevant to recycling and economically recyclable)

In consideration of the holistic philosophy already described, these targets cannot be based on recycling standards alone, but rather on overall ecological requirements focusing inter alia on the objectives in recycling. The RDC makes an important contribution in identifying internal conflicts of interest in the overall development of a product in good time, thus opening up the gateway for possible solutions.

The methods applied in dismantling analyses were developed on the basis of BMW's recycling standard, taking both ecological and economical criteria into account. To simplify the implementation and evaluation of dismantling analyses, the RDC has developed a special software tool called DAISY (**D**ismantling **A**nalysis **I**nformation **S**ystem). In the meantime other manufacturers and research institutes have adopted these analytical methods for their own purposes.

Detailed information on the methods and software used is to be found in [3].

The RDC has consistently developed and enhanced the DAISY software tool in the course of time, and is therefore in a position today to conduct virtual dismantling analyses. This, in turn, allows the RDC to perform another essential task in the Product Evolution Process, the supervision of predetermined targets: Once the first models of a new car are available, the RDC conducts a real-life dismantling analysis providing important recycling data long before the very first model is delivered to the customer.

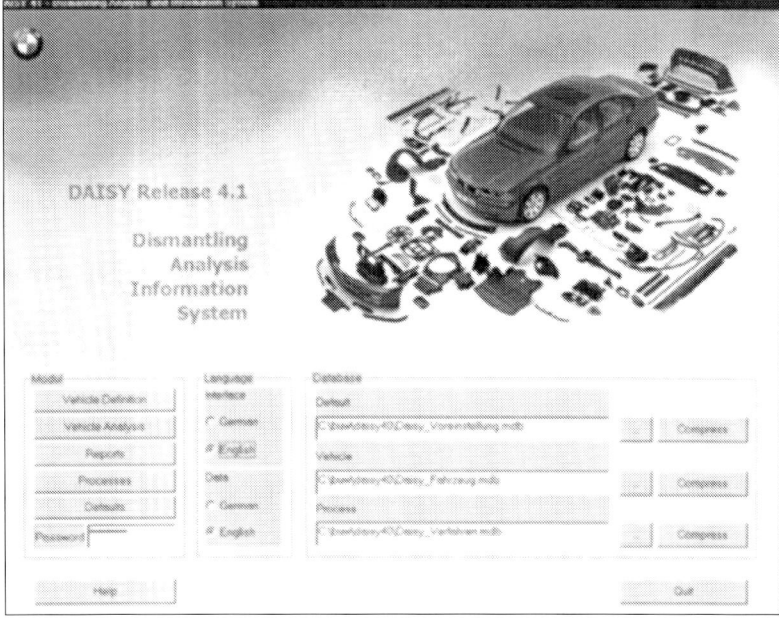

Fig. 5: DAISY dismantling software

To support BMW's technical departments and suppliers, the BMW Group's recycling specialists have compiled a Manual on Recycling-Optimised Product Design based on the know-how gained so far in the process of dismantling analysis. This Manual presents all important information for the recycling-optimised development of components and gives the designer/construction engineers suitable assistance in developing a component from the perspective of recycling.

Within this process the RDC acts as an intermediary between the recycling of a vehicle in future and the ongoing process of product development. Requirements resulting from practical recycling procedures are therefore integrated from the start into the process of product development. At the same time the development of special tools for vehicle recycling is determined from the beginning in the product development phase and may be initiated well in advance.

4.2.2 Life Cycle Assessment (LCA)

The responsible specialists consider not only the methods applied in DfR, but also possible ecological conflicts of interest. This they do by means of a standardised tool in environmental management, the so-called Life Cycle Assessment (LCA). In interdisciplinary discussions, alternative components are therefore assessed also in terms of their environmental compatibility and qualities for recycling, with the objective to achieve an optimum balance of both criteria.

The LCA serving to improve the environmental compatibility of products and systems is applied specifically in the early phase of product development in order to judge the environmental effects and repercussions of new concepts or technologies right from the start. Inter alia, this tool serves to ensure the fulfilment of environmental standards and political objectives. One example in this context is the voluntary commitment by ACEA to reduce passenger car CO_2 emissions by 25 per cent up to the year 2005 (proceeding from 1990 as the reference year).

The objective is to identify the most appropriate development option offering the greatest potential from a lifecycle perspective to enhance environmental compatibility. Obtaining a good foundation for ecological comparisons by way of interdisciplinary decisions taken by the development team as a whole is an important factor also in sensitising all parties involved to the growing environmental demands made of new vehicle generations.

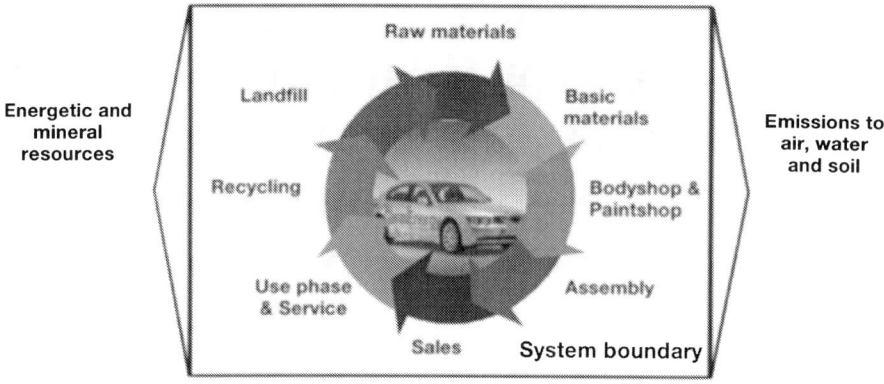

Fig. 6: LCA as specified in ISO 1404X as a tool supporting decisions in the Product Evolution Process

D002/005 © IMechE 2002

5. CONFLICTS OF INTEREST IN ENVIRONMENTALLY-ORIENTED VEHICLE DEVELOPMENT – CARBON FIBRE AS AN EXAMPLE

Under the European Directive for the Recycling of End-of-Life Vehicles passed in the year 2000, 95 per cent by weight of an end-of-life vehicle must be recovered as of the year 2015. The Directive furthermore calls for a level of material recycling exceeding 85 per cent by weight, only a maximum of 10 per cent by weight of the inevitable residual materials being used for the energy recovery.

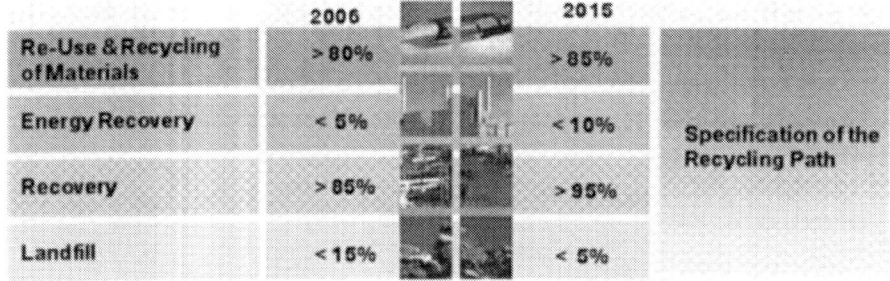

Fig. 7: Recycling and recovery quotas required by law under the EU ELV Directive

With increasing demands being made of the automobile in terms of its environmental compatibility, we find a fundamental conflict of interests between a high level of recycling and the minimisation of environmental effects such as CO_2 emissions per kilometre. Factors relevant to the environment such as CO_2 emissions depend directly on fuel consumption dictated to a large extent by the weight of the vehicle.

Lightweight construction using composite fibre plastics can reduce fuel consumption significantly. But such composite materials are largely regarded today as not economically recyclable. A further point often taken into account in comparisons with steel is the greater ecological burden involved in the production of such materials.

The question is therefore which system of materials reduces the burden on the environment most throughout the entire lifecycle of the product. Clearly, this point must be considered not just specifically for composite fibre plastics, but rather for all alternatives in lightweight construction [5].

Presenting the Z22 research car as a spearhead in technology, BMW Technik GmbH has shown the potential a manufacturer can achieve using a bodyshell made of Carbon-Fibre-Reinforced Polymer (CFRP). Not least through the extremely low weight of such a vehicle, average fuel consumption can be reduced to just 6 litres of petrol on 100 km (47.1 mpg Imp), even though the car has the dynamic performance of a BMW 528i touring.

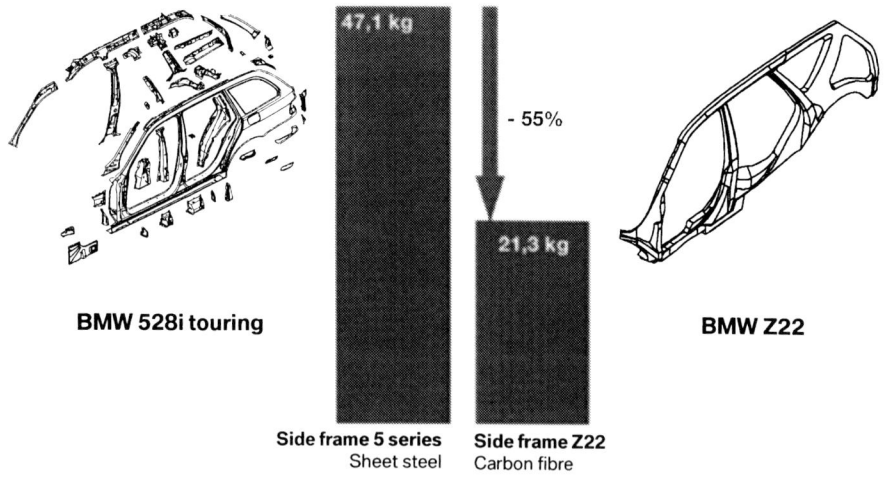

BMW 528i touring

47,1 kg

- 55%

21,3 kg

BMW Z22

Side frame 5 series
Sheet steel

Side frame Z22
Carbon fibre

Fig. 8: Weight saved by the use of CFRP for structural components

A method regularly applied in the automotive industry to judge products from an interdisciplinary perspective is the Life Cycle Assessment of components. Taking the example of the Z22 side-frame in deciding on whether to use steel or CFRP, BMW Group engineers have determined that the latter is able to provide significant positive effects in improving ecological standards. Primarily, of course, these effects result from the reduction of fuel consumption.

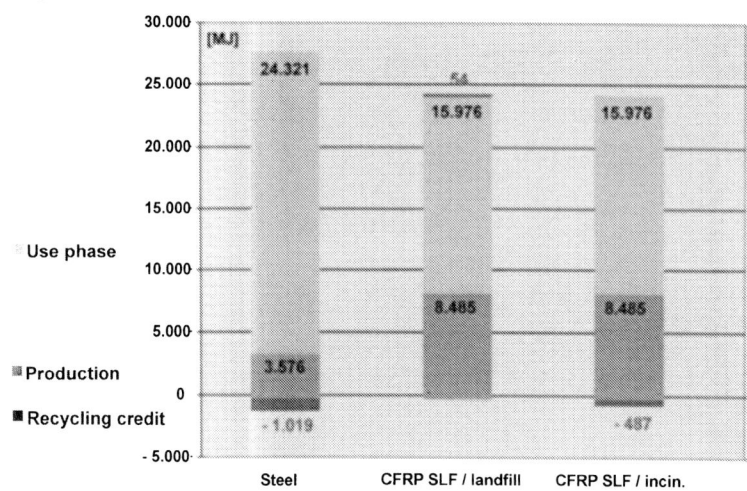

Fig. 9: Energy consumption over the lifecycle of the side-frame on the 5 Series touring (steel) and Z22 (CFRP)

D002/005 © IMechE 2002

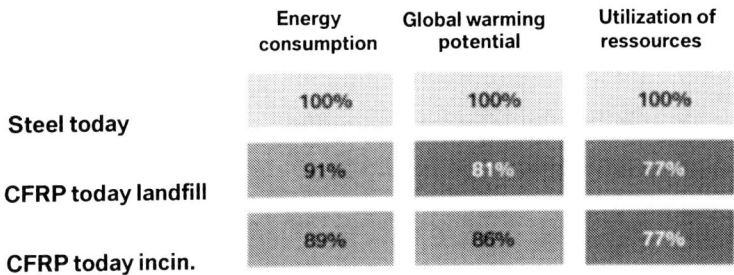

	Energy consumption	Global warming potential	Utilization of ressources
Steel today	100%	100%	100%
CFRP today landfill	91%	81%	77%
CFRP today incin.	89%	86%	77%

Fig. 10: Result of the impact assessment on a steel and CFRP side-frame

The disadvantage from today's perspective speaking against the regular use of CFRP in body construction is the lower recycling quota of the car possibly to be expected, since CFRP structural components are currently not regarded as economically recyclable. This drawback results not just from the material as such, but also from the position of the components within the car, with the dismantling process taking up a lot of time and therefore costing extra money.

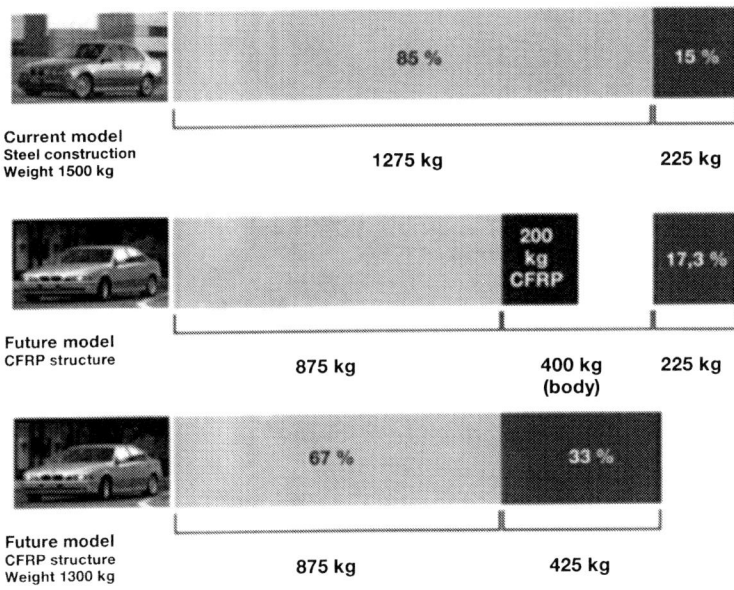

Current model
Steel construction
Weight 1500 kg

85 %   15 %

1275 kg   225 kg

Future model
CFRP structure

200 kg CFRP   17,3 %

875 kg   400 kg (body)   225 kg

Future model
CFRP structure
Weight 1300 kg

67 %   33 %

875 kg   425 kg

Fig. 11: Recycling quota of a hypothetical CFRP successor model to a current car with a recycling quota of 85 per cent

The recycling standard required by European legislation with a recovery quota of up to 95 per cent as of the year 2015 (85 per cent material recycling) thus presents a substantial barrier to such an ecologically sensible alternative. Greater use of regenerating raw materials also suffers from the limits imposed by law on energy recovery.

This is just one of many examples in the context of environmental legislation leading to the question of whether the focus in politics and legislation should in future be more on lifecycle thinking, or whether the conventional approach of thinking in specific problem areas and responsibilities will prevent a broader philosophy one might refer to as "integrated, all-inclusive" legislation.

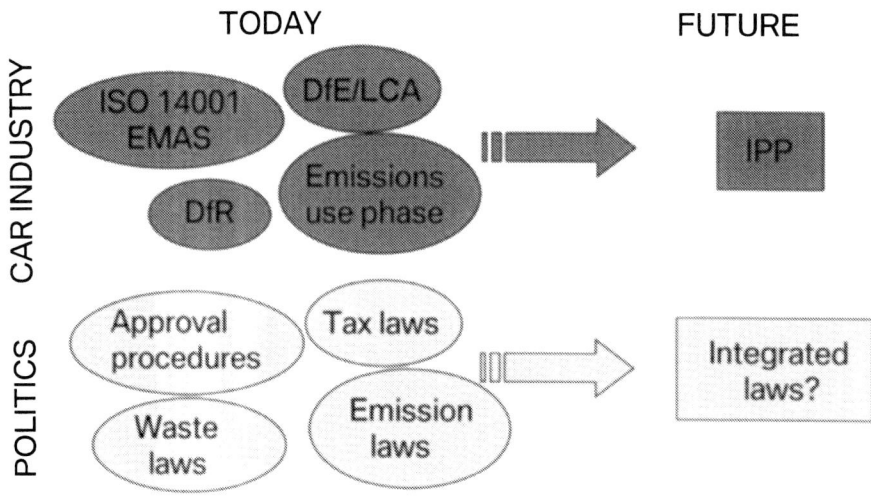

Fig. 12: Paths leading to an Integrated Product Policy

CONCLUSION

The article shows that the BMW Group readily meets the challenges presented by the principles of sustainable development. Precisely this is why the BMW Group has developed a broad range of tools for ecological product optimisation integrated and applied at crucial points within the product development process. A further point is that environmentally-oriented product development is consistently geared to environmental requirements at BMW's production plants and in recycling. This provides a sound and strong foundation for the "ecological cornerstone" of sustainable development with the BMW Group.

Taking CFRP as an example, the article also shows that external influences coming in particular from European and national legislation are often counter-productive to such efforts focusing on the ecology. In the case presented, strict maintenance of rigid quotas for 85 per cent material recycling obstructs or perhaps even excludes the use of an innovative material beneficial to the environment.

D002/005 © IMechE 2002

This is precisely where politicians, like other parties, must re-think their philosophy in the interest of "integrated" legislation not limiting, but rather enlarging the space for innovative solutions. Only joint efforts by industry, politicians and consumers will pave the way for sustainable improvements leading up to the desired standard of environmentally-compatible mobility.

[1]   N N: Umweltleitlinien des BMW Konzerns – Verantwortung für unsere Umwelt (The BMW Group's Environmental Guidelines – Responsibility for Our Environment). BMW AG, Munich.

[2]   N N: BMW Recycling – Recyclingoptimierte Produktgestaltung (BMW Recycling – Recycling-Optimised Product Design). BMW AG, Munich.

[3]   Zettier, T; Essenpreis, M; Vornberger, K: Evaluation of the Recyclability of Vehicles During the Product Development Phases. Society of Automotive Engineers, Inc., 2000-01-1469.

[4]   Essenpreis, M; Graser, K; Hagen, T; Pfleger, C; Vorberger, K; Wörle, G: Umweltverträglichkeit der neuen 7er-Baureihe (Environmental Compatibility of the New BMW 7 Series). Special Edition of ATZ and MTZ – The New BMW 7 Series. Friedr. Vieweg & Sohn Verlagsgesellschaft, Wiesbaden, Germany, November 2001.

[5]   Hagen, T; Vornberger, K: Mögliche Auswirkungen der Recyclingquoten und der ökologischen Produktbewertung auf die Automobilentwicklung (Possible Effects of Recycling Quotas and Ecological Product Assessment on Automobile Development). Proceedings of the 2001 Dresden Lightweight Construction Symposium, Dresden 2001.

Sustainability in fast-moving consumer goods

M SHAW
Safety and Environmental Assurance Centre, Unilever, UK

ABSTRACT

Fast moving consumer goods (FMCG) present a unique challenge with regard to minimising their environmental impact. Apart from the packaging, which can potentially be recycled, the components are designed to be used (for nutrition, to clean, groom etc) and then disposed of to waste. How can the environmental impact of such products be minimised so as to ensure their sustainable use?

The key is the application of Life Cycle Analysis (LCA) to understand where the major impacts of products are from cradle to grave (or cradle to cradle in a true sustainability sense) and to incorporate this understanding into product design and supply.

In Unilever we recognise the need for good stewardship in the use of resources

(i)   to utilise ingredients which are safe in use and subsequently degrade to have no harmful environmental impact and
(ii)   to continuously reduce the environmental of our manufacturing operations. This is eco-efficiency applied to the consumption of materials, energy and water and emissions to land, water and air. In all aspects we seek to continuously improve performance.

However the major environmental impacts of our products are in the raw material supply chain (agriculture for foods) and in use and disposal (home and personal care products). These impacts cannot be reduced by business operating alone but only in partnership with key stakeholders. These include suppliers, consumers and key opinion formers. The principle of partnership underpins the three major initiatives:
Sustainable Agriculture
Sustainable Fisheries
Sustainable Water
which drive the way to sustainable operations. They all include equally important aspects of corporate social responsibility.

In addition, environmental impact considerations are becoming essential criteria with product innovation. "Design for Excellence" is a framework which integrates, *inter alia*, functionality, delighting consumers and environmental improvements into winning product design.

1. INTRODUCTION

Conventionally, product design for the environment is predicated on the environmental impact being clearly associated primarily with the product itself throughout its value chain. That is from raw materials, through manufacturing, distribution, use and finally in disposal/reuse/recycling. Fast moving consumer goods (FMCG) – products brought from the supermarket and used everyday – on the other hand, present a unique challenge with regard to minimising environmental impact. The packaging components of such products can indeed be designed for both functionality (security, safety, ease of use etc) and re-use/recycling (where suitable schemes are in place, e.g. for glass bottles or metal cans). The functional product components, in contrast, are designed to be used – for nutrition, to clean and groom etc – and then be disposed of to waste. Also there is a large environmental impact, e.g. in energy and water consumption, associated with the use phase. In addition, the user and local infrastructures play a major role on the impact on disposal.

How can the environmental impact of such products be minimised to ensure their sustainable use?

2. THE UNILEVER APPROACH

The most effective management tool is the application of Life Cycle Analysis (LCA). Usually it is used to understand where the major impacts of products are from "cradle to grave" (*or from "cradle to cradle" in a truly sustainable sense*) and to incorporate this understanding into product design and supply. In addition, and powerfully, the methodology can be extended to an analysis of the total impact of the operations of a business – Overall Business Impact Assessment. Results from this can be used to determine the strategic thrusts which drive environmental improvement. This analysis applied to the Unilever operations (a €53 billion business operating in more than 100 countries) shows that potential impacts in nutrification and photochemical smog formation need to be addressed.

D002/012 © IMechE 2002

Unilever's Environmental Imprint

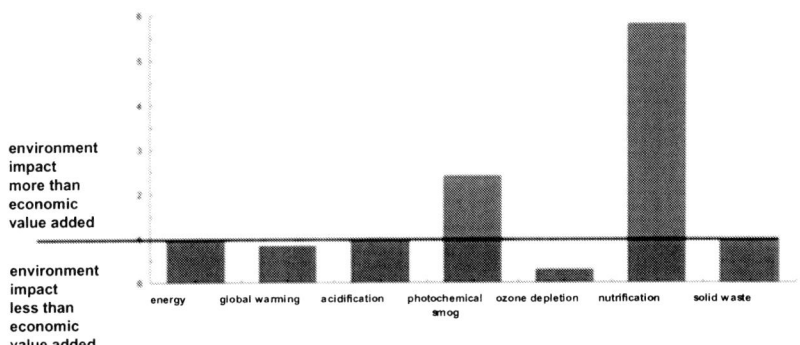

environment
impact
more than
economic
value added

environment
impact
less than
economic
value added

energy global warming acidification photochemical ozone depletion nutrification solid waste
                                        smog

Figure 1

2.1 Policy

This is the guiding element of the framework which recognises the need for good stewardship in the use of resources as an integral part of responsible corporate behaviour. Unilever's Corporate Purpose states "we believe that to succeed requires the highest standards of corporate behaviour towards our employees, customers and the societies and world in which we live". In our Code of Business Principles – which define how we conduct our business – is the following regarding the environment:

> "Unilever is committed to making continuous improvements in the management of our environmental impact and to the longer term goal of developing a sustainable business. Unilever will work in partnership with others to promote environmental care, increase understanding of environmental issues and disseminate good practice".

We put this into action mainly through 4 programmes:

Safe Ingredients
Responsible Manufacturing
Partnership Projects
Product Design

3.   SAFE INGREDIENTS

We select and utilise ingredients which are safe in use and which degrade to have no harmful environmental impact.   We do this based on an understanding of sound environmental science together with the application of leading edge risk management tools.

The cross-industry project in Europe, HERA – Human and Environmental Risk Assessment – is a pro-active, voluntary, response to address concerns about chemicals management. A series of raw materials consortia is producing dossiers – open and transparent - which make public the parameters used by the detergent industry to establish the requirements for safe use. The protocols have been endorsed by leading external stakeholders. 75%, by volume, of the materials used by the detergent industry in Europe will be covered by end 2002.

The principles of the approach will form an important element in industry's response to the EU chemicals management programme REACH (registration, evaluation and authorisation of chemicals).

4.   RESPONSIBLE MANUFACTURING

Life cycle analysis studies show, in general , that the environmental impact of our manufacturing operations is small compared to other impacts in our products life cycles. Nonetheless this is an area where responsible corporate behaviour tangibly makes good business sense.   Continually reducing the environmental impact of our manufacturing operations not only enables more sustainable use of resources and improves the environment but also contributes directly to the "bottom-line".

This is widely referred to as "Eco-efficiency" and we annually measure, report and set targets for six key parameters which pertain to the consumption of materials, energy and water and emissions to land, water and air.   The results are reported publicly on our web site www.unilever.com.

Unilever Environmental Performance 2000

Reduction in Load/Tonne of Production - expressed as % of 1996 value

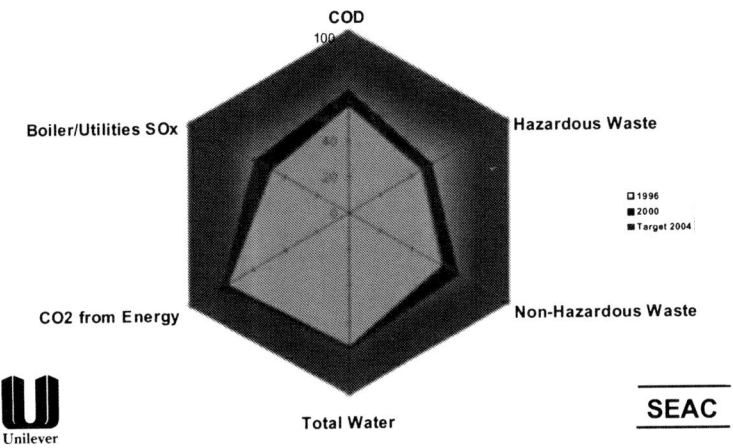

Figure 2

They show impressive progress. A major factor in this continuous improvement is ownership. Targets are set "bottom-up" from each manufacturing site (over 400) and reviewed annually at Board level.

There are processes in place which enable sites to freely share good practice. This is done within an effective framework for managing the environment. The Unilever Environmental Management System – fully based on the ISO 14000 standard – is mandatory world-wide. It has enabled 24% of our facilities to attain full ISO 14000 accreditation.

5.   PARTNERSHIP PROJECTS

Life cycle analysis shows that the major environmental impact of our products lies not in the products themselves but in the raw material supply chain (agriculture for foods) and in use and disposal (for home and personal care).

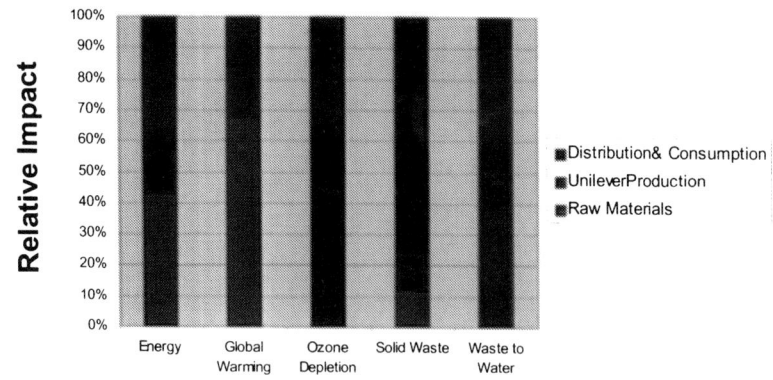

Figure 3

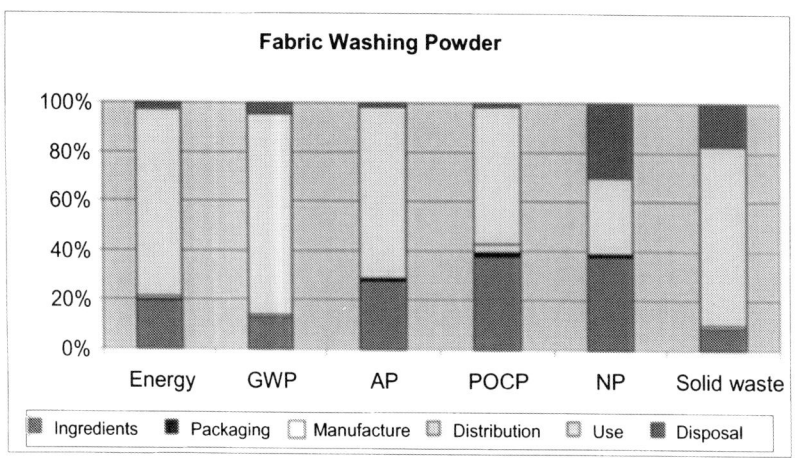

Figure 4

In addition we are one of the world's leading users of fish. We have a clear commercial interest in protecting and preserving fish stocks. How can we address this?

Finally, although Unilever is not in the "water" business, the availability of water matters to our business because it is essential in the production of raw materials and products and in consumer use. Unilever's imprint is associated with 0.1 % of the global fresh water availability.

D002/012 © IMechE 2002

To make progress in these areas of environmental concern we cannot operate alone – but only in partnership with key stakeholders. These include suppliers, consumers and key opinion formers. The principle of partnership underpins three major initiatives:

Sustainable Agriculture
Sustainable Fisheries
Sustainable Water

which are flagship projects leading the way to sustainable operations. All these projects integrate the triple bottom line aspects – economic, environmental and social – of sustainable development.

5.1 Sustainable agriculture

More than two thirds of the raw materials used in Unilever come from agricultural crops and livestock, fisheries and other potentially renewable sources. We are among the world's largest users of certain agricultural raw materials such as tea, vegetables, and vegetable oils. Unilever has shared the benefits of nearly half a century of dramatic increases in agricultural productivity brought about by scientific advances in farming, and the work and ingenuity of farmers throughout the world.

Despite huge successes over the past decades, all is not necessarily well with agriculture today. Experts are concerned, among other things, about a decline in soil fertility, soil loss through erosion, the availability of clean water, loss of biodiversity and the quality of rural life.

Environmental life cycle analysis studies have been carried out on Unilever's key food product categories including margarines, tomato-based sauces, frozen vegetables and black leaf tea. These studies assess the potential environmental impact of product systems across their whole life cycle from primary raw material production, agriculture, processing in a food factory, distribution and final consumption. For many of our food product systems the agricultural stage of the life cycle represents a significant source of potential environmental impacts.

An increasing world population and greater affluence means rising consumption and growing demands on the productivity of the soil. The associated increase in the use of machinery and inputs such as fertilisers, pesticides and fossil fuels will add to the burden on the environment.

Since the mid-1990s, Unilever has been consulting with experts and engaging with suppliers, customers, consumers and business partners around the world to find a sustainable way forward for agriculture. This has led to the following definition of sustainable agriculture:

"Sustainable agriculture is productive, competitive and efficient whilst at the same time protecting and improving the natural environment and conditions of the local communities."

This is why we support the following principles:

• Producing crops with high yield and nutritional quality to meet existing and future needs, while keeping resource inputs as low as possible.

- Ensuring that any adverse effects on soil fertility, water and air quality and biodiversity from agricultural activities are minimised and positive contributions are made where possible.
- Optimising the use of renewable resources while minimising the use of non-renewable resources.
- Sustainable agriculture should enable local communities to protect and improve their well-being and environments.

Unilever has initiated projects around the world to implement 10 indicators as a way to learn more about sustainable agriculture. We hope to understand and agree the ecological, social and economic conditions that sustainable agriculture must meet. This will eventually contribute to the development of standards for sustainable agriculture which we will share.

Pilot projects in tea, tomatoes, palm oil, peas, spinach and rape seed are assessing operations according to the following indicators:

1. Soil fertility/health
2. Soil loss
3. Nutrients
4. Pest Management
5. Biodiversity
6. Product value
7. Energy
8. Water
9. Social/human capital
10. Local economy

Our approach is to focus on the underlying health and vitality of agricultural systems – in social, economic and environmental terms. We believe that there needs to be a greater diversity of approaches to farm and plantation management. All agricultural systems have something to offer and we want to find out what works best under differing circumstances.

We are at the beginning of a very long journey and we need the continued help of others if we are to make progress. We are committed to engage with our stakeholders and we urge them to make contact if they wish to contribute in any way.

5.2   Sustainable fisheries

Unilever is one of the world's leading users of fish. We mostly use frozen white fish – cod, hake, hoki, Alaskan pollack and saithe. These are sourced from a number of fisheries in the Atlantic and Pacific oceans. Without regular supplies we would have no fish business. We have a clear commercial interest in protecting and preserving fish stocks, and we believe that consumers and the market can play an important role in fostering a sustainable fishing industry.

We are committed to buying all our fish from sustainable sources by 2005, and are working with suppliers towards meeting this target through a phased process of continuous improvement. We have invited suppliers to observe a code developed by the German fish industry that promotes sustainable fishing, and to date some 90% of our suppliers have signed this code.

D002/012 © IMechE 2002

We also work in partnership with the Marine Stewardship Council (MSC) to encourage suppliers to move towards certification to MSC standards. Filegro, a brand now using Alaskan salmon as an ingredient, was our first product to come from an MSC-certified fishery and we are now using New Zealand hoki in Europe, from fisheries certified to be well managed.

The Marine Stewardship Council (MSC) is an independent, global, non-profit organisation based in London. In a bid to reverse the continued decline in the world's fisheries, the MSC is seeking to harness consumer purchasing power to generate change and promote environmentally responsible stewardship of the world's most important renewable food source.

The MSC has developed an environmental standard for sustainable and well-managed fisheries. It uses a product label to reward environmentally responsible fishery management and practices. Consumers, concerned about overfishing and its environmental and social consequences will increasingly be able to choose seafood products which have been independently assessed against The MSC Standard and labelled to prove it. This will assure them that the product has not contributed to the environmental problem of overfishing.

Though operating independently since 1999, the MSC was first established by Unilever and WWF, the international conservation organisation, in 1997. It is this exciting and unique green-business partnership that has been praised by world leaders.

The MSC has succeeded in bringing together a broad coalition of supporters from over 100 organisations from over 20 countries.

Three principles underpin the work of the Marine Stewardship Council.

PRINCIPLE 1

A fishery must be conducted in a manner that does not lead to over-fishing or depletion of the exploited populations and, for those populations that are depleted, the fishery must be conducted in a manner that demonstrably leads to their recovery.

PRINCIPLE 2

Fishing operations should allow for the maintenance of the structure, productivity, function and diversity of the ecosystem (including habitat and associated dependent and ecologically related species) on which the fishery depends.

PRINCIPLE 3:

The fishery is subject to an effective management system that respects local, national and international laws and standards and incorporates institutional and operational frameworks that require use of the resource to be responsible and sustainable.

5.3　Sustainable water

The world's water systems – a shared resource – are under intense pressure. The availability of clean water matters to us because consumers need it to use our products and we need it to make them.

Three-quarters of our raw materials come from plantations or farms. Everything that is planted there – from tea to palm oil, spinach to peas to tomatoes – needs water to grow. Without adequate water supplies we would not have any raw materials.

Our factories need water to process our products – for washing, cleaning, cooling and making steam. Water is also a vital ingredient in many of our products and must therefore be clean and safe.

We all need clean water to make our tea, soup or sauce, to have a bath or to clean our teeth. Without clean water many of our branded products would be unusable.

Water is part of everything we do as a business – not just an arm's length intellectual exercise. We are involved at first hand in countries facing severe water stress and recognise that our activities have to be compatible with sustainable use of water wherever we operate.

Our starting point for understanding the role we can play in helping to ensure secure, sustainable supplies of water is to understand our own impact on water use – our 'water imprint'.

We have looked at our water use through the full life cycle of our products, and right across our product range, from raw material sourcing to consumer use of our products. This has given us a global picture of our water needs and of the way in which we impact on the availability of water.

% Annual Water Use By Life Cycle Stage
Total Unilever

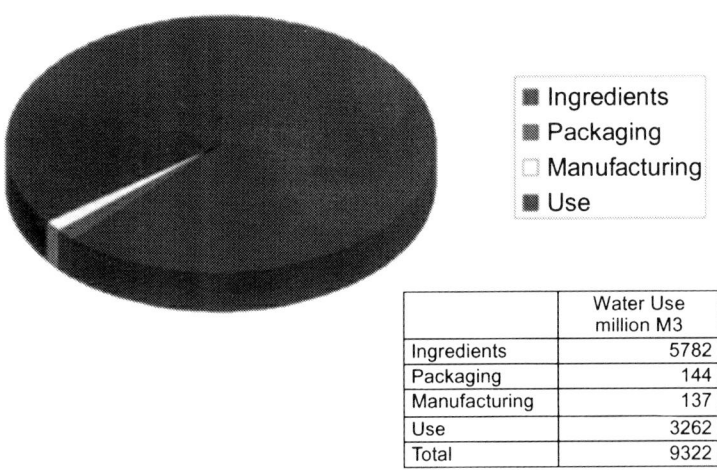

- Ingredients
- Packaging
- Manufacturing
- Use

	Water Use million M3
Ingredients	5782
Packaging	144
Manufacturing	137
Use	3262
Total	9322

Figure 5

We estimate that the total volume of water used to make and use our products is equivalent to around 0.1% of all the water extracted for use globally each year.

Our biggest user of water is agriculture. We estimate that potentially half of our overall water imprint must be associated with growing the produce that accounts for three-quarters of our raw materials.

We recognised that a global imprint is of limited value in guiding our action. We must get a better picture of the imprint of our local operations – particularly in those countries and regions already experiencing water stress. In addition, we wanted to work on the basis of a clear vision of the water question. This, we articulated as follows:

> To ensure that our activities and those of our suppliers, customers and consumers achieve a sustainable balance between protecting ecosystems and meeting human needs, so assuring the ability of future generations to access sufficient quantities of clean water. We will do this by understanding the water imprint of our operations locally and by ensuring that our imprint is sustainable within the limits of the relevant water catchments.

To this end we have set ourselves six objectives:
- Deepen our understanding of Unilever's impact on water resources by looking at the regional differences in our Water Imprint.

- Improve continuously water management in our factories to ensure that we minimise the contamination of water. In the past 5 years we have cut water pollution loading overall from our factories by 20 per cent. Many of our factories, particularly those in developing countries, discharge no effluent as a result of investment in on-site waste treatment and water recycling facilities.

- Help others, particularly our suppliers and customers, to do the same.

- Work in partnership with others to protect water catchment areas around the world.

- Contribute to finding effective solutions by sharing knowledge and best practice across our societies. We have formed an important link between our own scientists and the Institute of Water Research at Rhodes University, South Africa to work on aquatic toxicology.

- Promote "water awareness" and action on water by informing the public about ways to reduce water use and minimise waste water disposal. In Europe the detergent industry's "washright" initiative is demonstrating how effective advertising can be used to promote change in consumer behaviour for the benefit of water and the environment.

We believe that the key to sustainable water use lies, above all, at the local level. Our operating companies, deeply integrated in the societies in which they operate, are in the best position to explore local needs and to see how these needs can be met in the longer term.

Co-operation with local partners, is then, clearly critical. In October 2001, we published the results of a three year project to validate our approach to integrated water catchment management, known as SWIM. The initiative, was led by Lever Faberge here on Merseyside and supported by many local agencies including e.g., the Mersey Basin Campaign and the Environment Agency.

6.   PRODUCT DESIGN

"Green" products have been in vogue since the mid-80s. Despite much hype, and probably because of over-claiming by some manufacturers, market place success in our industry has not been high. We see that consumers obey the 30:3 rule. 30% say they purchase on environmental criteria. Analysis of supermarket trolleys show that only 3% actually do so and few niche products have found success. Does this mean that the "big brands" bought for everyday use are environmentally damaging? Of course all products have an impact on the environment but as a responsible business we seek to minimise this. Indeed by making continuous improvement to the environmental performance of big selling items, i.e. those which consumers prefer, our overall impact is substantially reduced. How do we integrate environmental considerations into product design? – "To add value while lowering impact".

"Design for the Environment" is a tool we have developed to strengthen the integration of environmental, economic and social factors early in the product design progress. In particular Design Space^{TM}

Design Space™

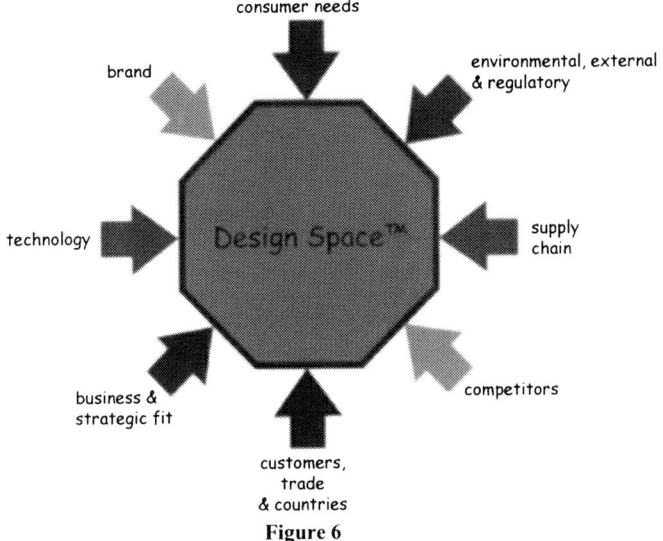

Figure 6

places eight elements needed for good product design around an equally balanced space. A successful innovation will have integrated the eight elements:
- consumer needs; environmental; external and regulatory; supply chain; competitor; customers, trade and countries; business and strategic fit; technology; brand.

This gives a framework of opportunity to integrate environmental and sustainability concerns into the business process. Some of these opportunities are:

- Sustainability: "How this brand does business"
- Ensuring all functions (including marketing) see the business/innovation potential
- Using understanding of critical impacts for each category to drive innovation
    - HPC impacts are primarily concerned with product usage
    - Foods impacts often concerned with supply
- Developing ways to engage consumers and promote product choice, through the right language and tone of voice
- Customers as communication partners to consumers
- Redefining the boundaries of competitiveness
- Building strong NGO network

Recent innovations in the Home & Personal Care Business demonstrate the effectiveness of this approach.

6.1   Laundry detergent tablets

Detergents tablets, introduced as a new format by Unilever in 1998 have been an enormous success and had a very positive impact on the environment. This format represents an excellent opportunity for our industry to make a major contribution towards the sustainable consumption of detergents. Tablet's ability to deliver improved environmental performance can be assessed in 3 areas:

- change of consumer behaviour
- less chemical disposal in the environment
- improved life cycle analysis indicators.

6.1.1   Positive change in consumer behaviour

Strong sustainability advantages happen when there is a good synergy between product benefits and evolving consumer habits. The beauty of the tablet format is that it satisfies the consumer need for convenience while, at the same time, offering more certainty in dosage for optimum results.

6.1.2   Less chemical material

Tablets enable controlled, lower dosage of detergents per wash. Controlled because of the format, lower because they inhibit the tendency to "add a bit extra".

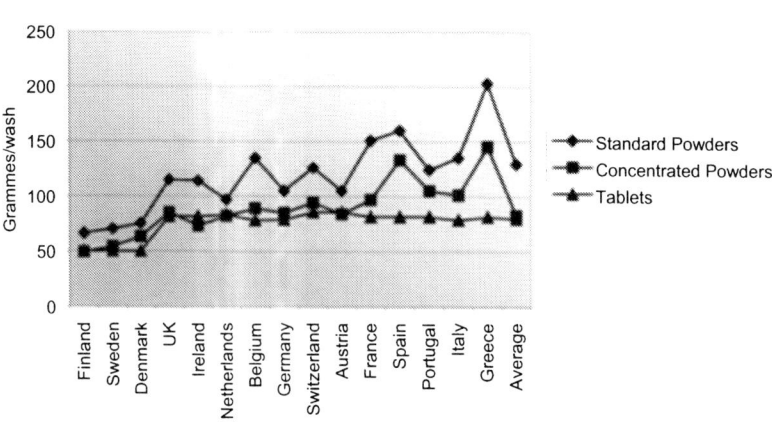

Figure 7

The immediate benefit of tablets has been a 53,000 tonnes reduction in chemical disposals to the environment in 1999 alone. Further improvements in resource utilisation has come from the packaging with a 26% reduction in packaging material per wash.

6.1.3   Life cycle analysis

Life cycle analysis methodology has been applied across a number of European countries to assess the effects of replacing the current major laundry detergent powders with tablets. The

                                               D002/012 © IMechE 2002

table below shows that a shift in use to tablets generates much more significant improvements in the selected environmental performance indicators than weaknesses.

Table 1

Significant Improvement	██████
Significant Worsening	██████
No Change	

Lever Tablets vs Lever Powders/ Country	Global Warming Potential	Acidification Potential	Photochemical Oxidant Creation Potential	Nutrification Potential	Solid Waste	Energy
Sweden				n/a*	██████	
Norway				▓▓▓▓		
UK			▓▓▓▓	██████		
Ireland				██████		
Netherlands						
Belgium		██████		██████		
Germany						
Switzerland		▓▓▓▓		▓▓▓▓		
Austria						
France		██████				
Spain		▓▓▓▓				██████
Portugal				▓▓▓▓		
Italy				▓▓▓▓		
Greece						██████

(*Not applicable as there are high levels of tertiary wastewater treatment systems throughout the country)

Other developments of Design Space are satisfied with this innovation. Primarily brands are strengthened but in addition there are advantages in the dialogue with regulators (supporting the industry "Code of Good Environmental Practice"), the supply chain is simplified (common product throughout the region), competitive advantage (market share) was enhanced, the business top priority category was strengthened and a new technology base established.

6.2   Other innovations

The successful methodology applied to Laundry Detergent Tablets has been applied to other product categories – for example - Machine Dishwashing Detergents and Liquid Laundry Detergents. In the former – the additional product and consumer benefit has been the development of "3 in 1" tablets. These combine the functionality of three conventional products – dishwasher detergent, water softening salt and rinse aid. The overall saving in chemicals and packing used per wash is dramatic.

Product usage reductions with 3-in-1's

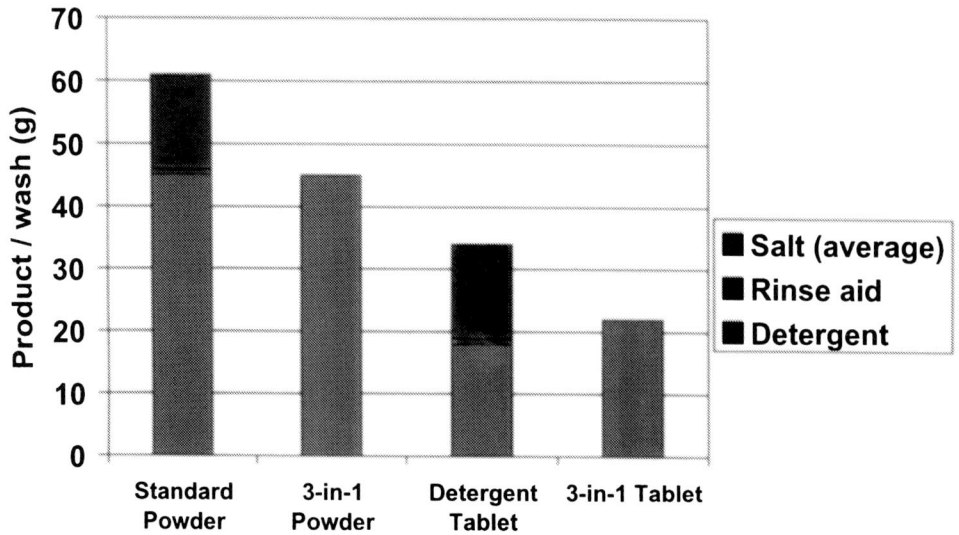

Product Format
Figure 8

Additional environmental benefits arise from transport savings. Transport requirements (no of 40 tonne trucks) have stayed consistent as the market has grown but the number of wash loads "delivered" has been increased by 60%.

The outcome of the life cycle analysis is similar as with laundry detergent tablets. Reflecting that nutrification was an important environmental issue to address we see substantial gains from the switch to tablets.

Table 2

Change from powder to tablets

COUNTRY	GLOBAL WARMING POTENTIAL	ACIDIFICATION POTENTIAL	PHOTOCHEMICAL OXIDANT CREATION POTENTIAL	NUTRIFICATION POTENTIAL	SOLID WASTE	ENERGY
Belgium						
France						
Germany						
Italy						
Netherlands						
Norway*						
Portugal						
Sweden*						
Switzerland*						
UK						

■ Improvement   ■ Worse   □ No Change   * Zero P formulation

Finally, laundry detergent liquids. Product innovation is one of the major influences in reducing the environmental impact of home and personal care products. Sustainable consumption of detergents has been promoted by linking it to other benefits which consumers find most important. The benefits derived from laundry detergent tablets have now been replicated with unit dose capsules for detergent liquids. A switch from the current "freely dosed" liquids to unit dose capsules will potentially save 50000 tonnes of chemicals and use up to 70% less packaging. The introduction of this new product form will contribute to the delivery of European Industry's "Code of Good Environmental Practice". This is a unique voluntary agreement with the EU community to substantial reductions in chemicals, packaging and energy when used in the clothes washing process.

Washright symbol. www.washright.com

Figure 9

7. CONCLUSIONS

Fast moving consumer goods – those that all of us use in great quantities everyday – have a large environmental footprint. Life cycle analysis shows that the majority of this footprint lies upstream and downstream in the product life cycle than the product itself. This understanding sets the targets for action. Improved product design will play a large part but only in partnership with suppliers and users with real progress be made in reducing environmental impact of these everyday products and facilitating their sustainable use.

Sustainable development – professional practice and systems thinking

A HALL
Environment Agency, UK
S MARTIN
Visiting Professor, Open University, UK

ABSTRACT

Professional bodies are now recognising that sustainable development is a key issue for professional practice and the wider role of professionals in society. Since many professional bodies also define the curricula of degree programmes, which provide the educational route to membership of the professions, this significant change in emphasis has far reaching implications for graduate and post graduate programmes in HE institutions. As part of this change process, 14 professional bodies have developed a common framework for sustainability to enable them to develop their thinking and practice. They have also engaged in a programme of inter-professional dialogue and learning to help develop a generic course on sustainable development for the professions, based on systems thinking and practice. The course explores the application of systemic thinking to a variety of organisational contexts that emphasises problem-solving, risk assessment and cultural change as part of the process to more sustainable businesses or organisations.

INTRODUCTION

Across the world, professionals and practitioners in a wide variety of public and private sector roles have begun to explore the opportunities and challenges of sustainable development. However, exploration is not action. Meaningful change is only just beginning. Inspite of the debate about 'the Next Industrial Revolution' we continue, by and large, to design, make, sell, use and dispose of the same products, with consumer expectation changing only very slowly. (This is reflected in the 30:3 syndrome – the phenomenon in which a third of consumers profess to care about companies' policies and record of social responsibility, but ethical products normally achieve less than 3% of the market).[1]

---

[1] *Who are the ethical consumers? (2000)* R Cowrie and S Williams Published by the Cooperative Bank

A significant number of professional bodies play a key role in defining the curricula of higher education programmes, which prepare students for a specific profession. Having phased out their own examinations, many professions now rely on 'accredited' degrees as the educational route to membership. We should not, therefore, underestimate the influence that professional bodies have on degree programmes and in turn the opportunity this presents in relation to sustainable development learning.

The Government's Sustainable Development Education Panel (DETR, 1999) has also set out a number of strategic goals for the professions. It recommends that by 2010 all professional bodies and industry lead bodies should have sustainable development criteria included within their course accreditation requirements.

Professional bodies are increasingly being asked to review their traditions and practice - radically and urgently - with far reaching implications for those HE courses for which they control or influence the curricula. The challenge of sustainable development has profound implications for professions across a range of disciplines - whether engineering, planning, chemical, environmental, accounting, design, manufacturing or whatever profession - in both the practice and role of the professional. Engineers, for example, in designing solutions to meet modern needs, are responsible not only for the safety, technical and economic performance of their activities, but they also have responsibilities to use resources sustainably; to minimise the environmental impact of projects, wastes and emissions; and to use their influence to ensure their work brings social benefits which are equitably distributed. These responsibilities heighten the importance of ethics in curriculum design and require greater emphasis on codes of conduct and the role of engineers as social change agents.

SUSTAINABILITY AND SUSTAINABLE DEVELOPMENT

The key driver for much of this change is the significant shift in policy in the UK and elsewhere, from a focus on the environment to the wider context of sustainable development. This shift began in earnest in 1992, following the Rio Earth Summit – sustainable development has become the buzzword of the past decade – and most people think it is 'a good idea'. The concept has come a long way and the critics who predicted it as a passing fad that would quickly fade into oblivion are being proved decisively wrong. *'With the possible exception of "democracy", there currently exists no more widely endorsed symbol for positive socio-economic and political change than sustainable development'*[2]. It will be interesting to see what progress has been made during the past decade following the Rio +10 World Summit in Johannesburg.

Despite general support for the idea of sustainability, sustainable development (with its 250 plus published definitions) continues to provoke controversy. We can all aspire to the end goal of sustainability, but the means to its achievement – sustainable development – continues to be debated. What are the principles of sustainable development? How can they be applied? What would a sustainable world look like? What must people do to achieve it? Are the goals of continuous economic growth and sustainable development compatible? How can there be sustainable use of non-renewable resources? Should the developed world reduce its standard of living (and does it need to) in order that there is a fairer distribution of resources

---

[2] *Towards Sustainable Development: On the Goals of Development – and the Conditions of Sustainability (1999)* ed WM Lafferty and O Langhelle   Macmillan Press

between countries? These are just a few of the hundreds of controversial social, economic and environmentally driven issues to be debated. There are no right or wrong answers, but progress towards achieving a sustainable future lies in our willingness to think about these issues and to work towards solutions. A group of professional institutions and partner organisations have started to do just that - with a view to helping themselves and others work towards solutions to their own particular problems – via the Professional Practice for Sustainable Development project.

PROFESSIONAL PRACTICE FOR SUSTAINABLE DEVELOPMENT

Professional institutions constitute a range of individuals whose beliefs and values towards sustainable development are mainly derived from their long education, training and experience in their basic discipline. These are reinforced through their professional networks. If there is to be a common approach for sustainable practice amongst professionals, then the framework and training for this needs to come through their professional bodies. The Professional Practice for Sustainable Development initiative, often referred to as PP4SD, arose out of this kind of thinking. Working with 14 professional institutions the project aims to help members improve their capacity to plan and carry out their professional duties in ways that support their achievement of sustainable development (1).

The project started in June 1999 with funding from the Environmental Action fund following an invitation seminar in March 1999 initiated and hosted by the Environment Agency and the Council for Environmental Education. Its specific objectives are:

- To engage the participating professions in a learning process to develop a common curriculum framework for sustainable development.
- To develop, test and publish training materials derived from the framework appropriate to the needs of the professional institutions.

It is hoped that more professional institutions decide to join the work of PP4SD and engage in its Forum (to be set up in 2002) for developing and sharing ideas, tools and experience.

PP4SD - TOOLS AND TECHNIQUES

The PP4SD project generated an 'evolving' framework for sustainability to enable participants to apply a shared mental model to think about sustainability based on non-prescriptive principles that could be applied over a range of activities. The framework is set in a future perspective and, therefore, offers a useful tool to help describe the gap between today's activities and the future requirements of a sustainable society. It was derived from a number of key sources, including: The Rio Declaration, World Business Council on Sustainable Development, the (then) Department for the Environment, Transport and the Regions, The Natural Step, The International Institute for Sustainable Development, the World Commission on Environment and Development, Forum for the Future and Natural Capitalism. Whilst acknowledged to be evolving it currently states that:

In a sustainable society:

1   *Any materials mined from the earth should not exceed the environment's capacity to disperse, absorb, recycle or otherwise neutralise their harmful effects to humans and the environment.*

2   *Synthetic substances in their manufacture and use should not exceed the environment's capacity to disperse, absorb, recycle or otherwise neutralise their harmful effects to humans or the environment.*

3   *The biological diversity and productivity of ecosystems should not be endangered.*

4   *A healthy economy should be maintained, which accurately represents the value of natural, human, social and manufactured capital.*

5   *Individual human skills, knowledge and health should be developed and deployed to optimum effect.*

6   *Social progress and justice should recognise the needs of everyone.*

7   *There must be equity for future generations.*

8   *Structures and institutions should promote stewardship of natural resources and the development of people.*

As far as possible any approach to sustainable development needs to encourage professionals to internalise the general principles set out in the PP4SD framework and to work out for themselves, the implications or applications, as they relate directly to their professional activities. In order to support this process the PP4SD project has published two booklets that respectively describe (Book 1) the project's objectives and the role of the professional institutions and partners, and (Book 2) general support in the development of training courses and associated materials and tools. The recently published generic foundation course[3], based on systems thinking, provides practical and thought provoking methods to enable participants to assess their own knowledge, skills and experiences in the context of sustainable development. Case studies from business and industry are used to illustrate how sustainable development principles are being applied and to provide an opportunity for participants to develop their own thinking around practical examples.

SYSTEMS THINKING (2)

So what is wrong with the linear approach – Cause → Effect → Stop? Generally speaking, this more traditional approach to problem solving is often short-term, usually reactive, does not consider the full implications of any actions taken and at worst is potentially damaging ... without getting to the root cause of the problem. An example of this was the use of catalytic converters in petrol-engined cars (an end of pipe solution). There was terrific pressure from environmental groups to cut harmful exhaust emissions, by fitting all petrol-engined cars with catalytic converters. An alternative solution – the lean burn engine – was available but would have taken time and money to develop. The short-term fix won the day, but if a systems approach had been applied to the problem, the solution would have been different, as other factors would have been considered.

A simple way of applying systems thinking is to make two assumptions when making plans:

---

[3] *Professional Partnerships for Sustainable Development (2001)* – J Baines, J Brannigan and S Martin
Institution of Environmental Sciences

D002/025 © IMechE 2002

- Everything affects everything else. Try to anticipate what those effects are.
- There is no such thing as a free lunch. Look for the potential costs when you think you have found the perfect solution – they may be some way into the future.

A systems thinking approach to education and training requires a change of emphasis from instruction to learning. A systems approach will:

- emphasise increased capacity for self-reliance, self-correction, self-direction, self-organisation and self-renewal in our educational environment
- include system change as a process of problem solving
- develop awareness of the whole and recognise that nothing is isolated in a system
- focus on cooperation rather than competition
- show everyone they are part of systems and contribute to them
- focus on long-term consequences and root causes and avoid taking the easy way out
- emphasise personal responsibility and encourage appropriate informed actions

CHALLENGE FOR THE FUTURE

Professionals from whatever background or discipline – including those in education and training – are in a pivotal position to effect positive and sustainable change for society in the future. The future starts now … how are **you** planning to contribute?

ENDNOTES

(1) The professional institutions involved in this phase of the project are: Building Services and Research Information Association, Chartered Institution of Building Services Engineers, Chartered Institution of Water and Environmental Management, Chartered Institute of Purchasing and Supply, Institute of Energy, Institute of Waste Management, Institute of Chemical Engineering, Institute of Civil Engineers, Institution of Environmental Sciences, Institute of Mechanical Engineering, Royal Institute of British Architecture, Royal Institute of Chartered Surveyors, Royal Society of Chemistry, Royal Town Planning Institute.
(2) Further details of Booklets 1 and 2 and the Sustainable Development Foundation Course can be downloaded from the Institution of Environmental Sciences website: http://www.ies-uk.org

References

Baines, J, Brannigan, J, Martin, S (2001) *Professional Partnerships for Sustainable Development.* Institution of Environmental Sciences, London.

Cowrie, R, Williams, S (2000). *Who are the ethical consumers?* Cooperative Bank.

DETR (1999) *Sustainable Development Education Panel - First Annual Report 1998.* DETR, London

HMSO (1993) *Environmental Responsibility - an agenda for further and higher education.* HMSO, London.

HMSO (1996) *Environmental Responsibility - a review of the 1993 Toyne Report.* HMSO, London.

Lafferty, W M, Langhelle, O (ed) (1999) *Towards Sustainable Development: On the Goals of Development – and the Conditions of Sustainability.* Macmillan Press

Annie Hall*
Head of Education
Environment Agency
annie.hall@environment-agency.gov.uk

Stephen Martin**
Independent Consultant and Visiting Professor at the Open University
esm@esmartin.demon.co.uk

* Annie Hall is Head of Education, Environment Agency England & Wales and is a member of the Government Panel for Sustainable Development Education. She is also a member of PP4SD project management group.

** Stephen Martin is a member of the PP4SD project management group and the Sigma project steering group and visiting Professor at the Open University Centre for Complexity and Change.

Overviews

Evaluation of effective improvement strategies and successful measures for sustainable product design

R ZÜST
Alliance for Global Sustainability, Swiss Federal Institute of Technology (ETHZ), Zurich, Switzerland
W WIMMER
Institut für Konstruktionslehre, Vienna University of Technology, Austria

ABSTRACT

"Sustainable product design" aims at improving the environmental, economic, and social performance of a product over all life cycle phases. This requires a specific way of life-cycle-thinking, a target-oriented search for effective improvement strategies, a selection of successful measures, and an efficient implementation of these measures in the ongoing planning, decision-making and management processes. The main objective of this contribution is to describe these four planning and decision-making steps, the practical application of the ECODESIGN PILOT (Product Investigation, Learning and Optimisation Tool for Environmentally Sound Products with CD-ROM) [1] in education and application as well as examples from industry which illustrate how this concept can be used in practical applications.

1 INTRODUCTION

1.1 Sustainable Solutions are in Demand
More and more people seem to be realising that the current global economy is not sustainable. New solutions are under order. The magic phrase of the moment is "sustainable development" and is defined as incorporating economic, ecological and social aspects in one common perspective. This calls for multidisciplinary thought processes and action as well as actual practical examples. This comprehensive approach is needed to secure the long-term success of companies. Needed are new, proactive approaches to solutions, especially in the area of individual responsibility.

1.2 What is "Sustainable Product Design"?
A product has various ecological, social, and economic implications in its lifetime from the raw material stage to manufacturing, distribution, product use and finally, disposal. Environmental effects are caused for example when we take resources directly out of the ecosphere or when we add emissions directly into the ecosphere. These effects can cause environmental problems such as smog, greenhouse warming potential, and over-fertilisation

of water depending on the time and place of their influence. Similar cause and effect relationships come into play in the economic and social areas.

The individual life phases must be optimised as a whole if useful solutions are to be achieved. Here Life-Cycle-Engineering (see for example [2]) or product-life-cycle-management play an important role. In addition, all auxiliary processes which are necessary for example for the optimal operation of products must be considered. An isolated product must be looked at as a product system which must be evaluated and optimised (also see [3] and [4]).

The goal of sustainable product design is primarily to reduce the negative effects of a product on the ecological, economic, and social subsystems as much as possible, while at the same time generating the greatest possible benefit.

1.3 Why "Sustainable Product Design"?
There are various reasons why to put sustainable product design into practice in one's own enterprise (see also [1]). One important incentive is better environmental performance of one's own products. This is possible for example by avoiding or reducing potentially negative environmental impact, by reducing material and energy intensity of a product's life cycle, as well as by fulfilling health and safety standards in the enterprise.

Environmental pollution is often related to resource consumption. This consumption often results in cash flow. When relating sustainable product design to better resource management, it can thus lead to more cost-effective structures within a company. New product ideas can also come out of such self-critical reflection on one's own products, be it through life-cycle-thinking or the search for effective improvement strategies or successful measures. New and innovative product ideas can develop through interdisciplinary working methods, the questioning of individual life stages, and the optimisation of the product taking into consideration the consumers' actual needs. A further reason to make use of sustainable product design is to secure the future success of the enterprise by realising one's responsibilities, motivating one's employees, as well as gaining the trust of one's stakeholders.

1.4 How to achieve the concept of "Sustainable Product Design"?
The successful implementation of sustainable product design requires:

- Life-cycle-thinking (definition of the significant impact of a product or a product idea),
- definition of effective improvement strategies,
- selection of successful improvement measures, and
- efficient implementation in the already existing planning, decision-making and management processes of the enterprise.

The procedure described corresponds with already known problem-solving concepts (for example [5], [6], [7]). The step-wise procedure – from an idea to concrete solutions – is also described by [8] in the context of concept synthesis and concept analysis in the Zurich "Systems Engineering" methodology [7]. ECODESIGN PILOT [1] is also based on an analogous procedure.

In the following contribution, there will be a description of the first four planning stages - based on the ECODESIGN PILOT - in greater detail and some simple examples from ecological product design as means of clarification. Different planning, evaluation and decision-making tools already exist in this area (for example [1] or the technical report about "Design for Environment" (ISO 14062) [9]).

This contribution is based on the following works:

- Wolfgang Wimmer: The ECODESIGN PILOT [1]
- Michael Frei [10], student projects ([11], [12]) and discussions within the AGS-project "Decision Support for Planning Eco-Effective Product Systems from 1998 - 2000"
- Gabriel Caduff [13] and [4] and the discussions within the standardisation process in ISO/TC 207, especially "Environmental Performance Evaluation" [14]
- basic information about problem solving methodologies ([5], [6], [7], [15], [16], [17]).

2 LIFE-CYCLE-THINKING

Product life cycles of new products have to be analysed and looked at for possible effects to the ecological, economic, and social environment during the developing phase. This is often quite difficult since the individual impact of a product is usually hard to determine and weight, especially when trying to examine effects related to future customer behaviour and product usage.

2.1 Product example "Flushing System"

The example described is from an enterprise in the field of sanitary products. Fig. 1 shows an environmental evaluation of a flushing system, which is used in Switzerland, Austria and Germany. For the quantitative evaluation already existing eco-data-basis (e.g. [18]) as well as specific evaluation methods (e.g. [19] and [20]) have been used.

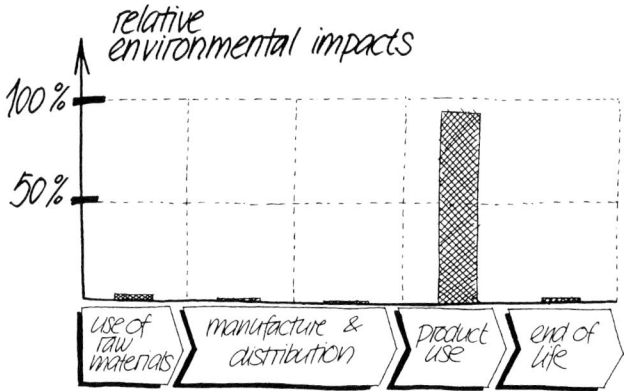

Fig. 1 Relative environmental impact of a flushing system (data-base [11], published in [21])

Additional water is needed to ensure that the cistern can fulfill its flushing function during the usage phase. One can calculate a certain number of flushing processes in the system's lifetime. Fig. 1 shows:

-   95% of the environmental impact of the whole life cycle occurs in the usage phase (especially the energy consumed for the transportation of water).
-   The environmental impact from use of raw materials, manufacturing, distribution, and disposal is not significant.

Based on the existing analysis, measures to reduce the environmental impact of the flushing system should be taken mainly in the usage phase.

2.2 Product example "Energy and Signal Cables"

The next example describes the relative environmental impact of various cable types with various usage scenarios. On the one hand, the study looks at stationary energy and signal cables in buildings. On the other hand, the examination focuses on cables that are moved in the usage phase, as in cars, railway wagons, or planes. The results are summarized in Fig. 2.

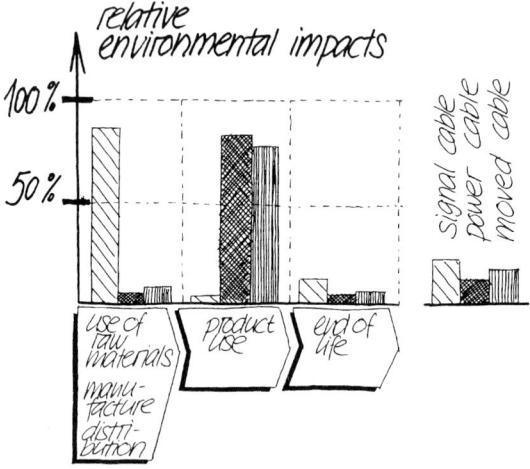

Fig. 2 Relative environmental impact of various cable types (data basis [12], published in [10])

Again, the results show significant differences:

-   Products such as signal cables cause significant environmental impact in use of raw materials, manufacturing and distribution. The power dissipation in the form of transmission loss is very low.
-   The opposite is true for energy cables. The power dissipation in the usage phase due to ohmic resistance is substantial. "Minimised material intensity" is an often described strategy in the context of ecological product design. The result in Figure 2 shows that

                                        D002/024 © IMechE 2002

this strategy would be wrong. The diameter of the energy cables should be larger to improve the overall environmental performance.
- The third scenario focuses on cables that are moved regularly in the usage phase as in cars. The cable in this case contributes to the total mass and is continuously accelerated and slowed down again by the vehicle. This large mass leads to high energy use during the usage phase.

2.3 Product Characterization as a Basis for Strategy Determination

The results of life-cycle-thinking are models which show all a product's life phases as well as the significant impact on the ecological, economic, and social environment. The examples in Fig. 1 and Fig. 2 show that different effects arise in the individual product life phases depending on the product and the usage scenario employed. Some of the products show substantial effects, i.e. a maximum is reached in a specific life phase. In the case of the flushing system this maximum in terms of environmental impact is reached in the usage phase.

One can define five different basic types of products with regard to the ecological, economic, and social implications. These can be described as [1]:

- "raw material intensive",
- "manufacture intensive",
- "transportation intensive",
- "use intensive"[1] and
- "disposal intensive" products[2].
- In addition, there are so-called hybrids with significant impact in two or more life phases

The breakdown into basic types is needed to convey the manner and characteristics of the examined product and to help find new solution paths that could be approached.

3 SELECTING IMPROVEMENT OBJECTIVES AND STRATEGIES

This section deals with the allocation of effective improvement strategies based on these product-specific characteristics.

3.1 Specific evaluation of improvement strategies

After the life-cycle-thinking step which consisted of examining the ecological, economical and social weak points of a product, the focus now turns to finding and defining effective improvement strategies. The following list shows various improvement strategies from an ecological point of view ([1]):

- Reducing material inputs
- Reducing energy consumption in production process

---

[1] The mass could be a significant aspect especially in the case of frequently moved objects (compare example with the different usage scenarios of energy and signal cables in the section before).

[2] A single product can have a local, regional or global distribution. This is the reason for different distribution, service, and recycling concepts. Another reason could be the fact that there is a change of ownership. Manipulations in the usage phase also have an influence on the recycling system. An efficient recycling or upgrading is no longer possible.

- Optimizing type and amount of process materials
- Avoiding waste in the production process
- Ecological procurement of external components
- Reduction of packing
- Reduction of transportation
- Optimizing product functionality
- Improving maintenance
- Ensuring environmental safety performance
- Reducing consumption at use stage
- Avoidance of waste at use stage
- Increasing product durability
- Improving reparability
- Improving disassembly
- Reuse of product parts
- Recycling of materials

The effectiveness of sets of strategies needs to be tested for each specific case. Working in an interdisciplinary team can ease this exercise. Additional tools and methods in the field of environmentally compatible product design can be found in the ECODESIGN PILOT [1] or on the website: www.ecodesign.at . As the following section illustrates with the example of the washing machine, different usage scenarios can also lead to varying improvement strategies.

Washing Machine A – Intensive Usage
Washing machine A is used intensively at approximately 10'000 washing cycles in its lifetime (Fig. 3).

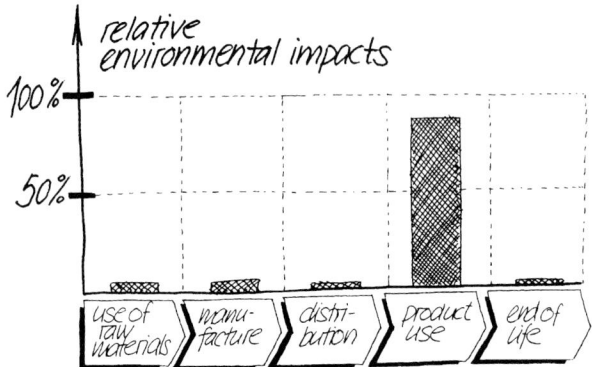

Fig. 3 Relative environmental impact of a frequently used washing machine

The greatest environmental impact in this case is in the usage phase. The strategies deemed as promising for an intensively used product are shown in the following list [1] or in the ECODESIGN online version on: www.ecodesign.at):

Optimizing product functionality

 D002/024

- Improving maintenance
- Ensuring environmental safety performance
- Reducing consumption at use stage
- Avoidance of waste at use stage

The discussion shows the following:

- "Improving maintenance" only makes sense if the washing process can be improved by regular monitoring. This is not the case for the example of the washing machine.
- "Ensuring environmental safety performance" is irrelevant here.
- Waste in the narrower sense of the word is not produced during the washing process. "Avoidance of waste at use stage" does not play a role in this example.
- "Optimizing product functionality" as well as "reducing consumption at use stage" are used to optimize the washing process in the actual process control as well as in the use of detergent. Both strategies are reasonable.

The two strategies "optimizing product functionality" and "reducing consumption at use stage" are now the basis for deducing concrete improvement measures.

Washing Machine B – Infrequent Usage

Washing machine B is situated in a small household and is infrequently used at approximately 1000 washing cycles in its lifetime (Fig. 4).

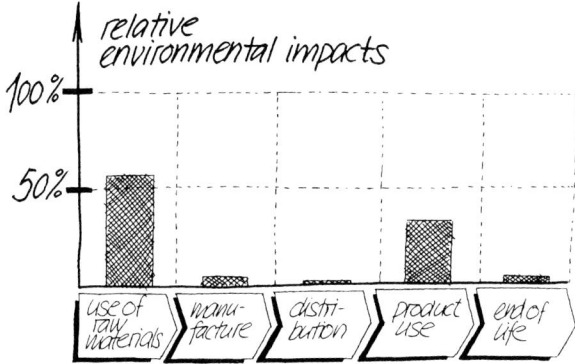

Fig. 4 Relative environmental impact of an infrequently used washing machine

This washing machine is an example of a hybrid that shows two environmentally related stress maximums. The improvements are focused on two areas, namely the differentiated design of the machine with respect to better and longer usage of the materials and the improved washing process.

The strategies deemed as promising for this hybrid are the following ([1] or on the ECODESIGN online version on: www.ecodesign.at):
- Selecting the right materials
- Reducing materials inputs

- Optimizing product functionality
- Ensuring environmental safety performance
- Reducing consumption at use stage
- Avoidance of waste at use stage
- Increasing product durability
- Improving maintenance
- Improving reparability
- Improving disassembly
- Reuse of product parts
- Recycling of materials

The discussion demonstrates the following:

- The strategies "optimizing product functionality" as well as "reducing consumption at use stage" are reasonable strategies (cf. washing machine A).
- The strategy of "reducing materials inputs" could lead to a simpler design in this case. Since the product is not used often, a less solid construction should not have many disadvantages.
- Aiming for a longer lifetime with the strategy of "increasing product durability" does not make sense in this case. The machine will be technologically outdated before the end of its lifetime. The strategy of "reuse of product parts" can be dismissed for the same reason.
- Washing machines consist of "raw material intensive" parts, e.g. the washing drum. The strategies "selecting the right materials", "improving disassembly", "reuse of product parts" and "recycling of materials" should be considered in this case.

As shown above, the set of strategies for washing machine B is different from the one for washing machine A.

3.2 Conclusion
The deliberation on improvement strategies is the first step in synthesis-analysis. The solution path is then roughly drafted using the chosen improvement strategies. At this stage it is important to focus on the basic essentials.

By explicit separation of the two planning steps, "effective improvement strategies" and "selection of successful measures", one is able to develop and eliminate solutions step by step. This reduces the means invested, while at the same time increasing the planning reliability. Developing and eliminating options step by step, in other words evaluating and selecting them, is a typical element of successful planning and problem-solving methodologies and is the basis of general product design.

4 SELECTION OF SUCCESSFUL MEASURES

The next step entails finding and choosing the measures within the selected strategies which will result in the greatest possible ecological gain at the lowest possible cost.

Each improvement strategy consists of a great number of potential measures as shown in the following example of the strategy in Fig. 5, "improving disassembly ".

D002/024 © IMechE 2002

The main points related to ecological product design are documented in the ECODESIGN PILOT [1]. In addition the ECODESIGN PILOT supports the selection of useful improvement measures by providing a set of improvement measures as well as tools and methods for the selection process.

The improvement measure "ensure easy access to connecting parts" for example influences not only the environmental performance, but also the methods for product repair and revision (social environment) as well as the economic environment. The question when selecting potential measures thus remains which option would result in a maximal ecological, social, and economic gain whilst using a minimal amount of means. It is sensible to approach such questions in interdisciplinary workshops or with the help of appropriate evaluation methods. The evaluation process of possible improvement measures in general is the same as in the phase of life-cycle-thinking. There is also a need to evaluate a specific situation in a wider sense, that would mean from an ecological, economic, and social point of view.

strategies:	*possible improvement measures:*
"Improving disassembly"	- Ensure easy access to connecting parts
	- Ensure reversibility of assembly procedure
	- Design product structure for easy disassembly (uniform directionality for assembly and disassembly work)
	- Minimise time and paths for disassembly
	- Use easily detachable connections
	- Ensure easily visible access to connections for disassembly
	- Ensure easy access to connecting parts for disassembling tools
	- Ensure functioning of connections over whole service life

Fig. 5 Potential improvement measures for "Improving disassembly" ([1] or ECODESIGN online version on: www.ecodesign.at):

5   THE ECODESIGN PILOT METHODOLOGY AND THE EFFICIENT IMPLEMENTATION OF IMPROVEMENT MEASURES

The ECODESIGN PILOT [1] provides planning and decision making methods and specific EGODESIGN information. Applying the ECODEISGN PILOT to specific products consists of a seven-step procedure, explained as follows:

1. Life Cycle Thinking: This is to identify those phases in the product service life that cause considerable environmental impact.
2. Finding product characteristics: The ECODESIGN PILOT characterises products by means of "basic types".
3. Selecting improvement objectives and strategies: Finding the appropriate strategy for a certain product type (see Fig. 6).
4. Identifying of ECODESIGN guidelines: Working with checklists to identify the appropriate ECODESIGN measures for the improvement of a product (see Fig. 6).

5. Transforming ECODESIGN guidelines into individual product design changes: The transformation from the generally formulated ECODESIGN guidelines into product-related measures requires experience and creativity.
6. Evaluating and carrying out design changes: The evaluation is a comparison of the potential environmental improvement and the potential risk to be expected from implementing certain measures.
7. Decision and organisation of the identified measures: Setting a time frame to carry out the measures and determine a person or department responsible for the realisation.

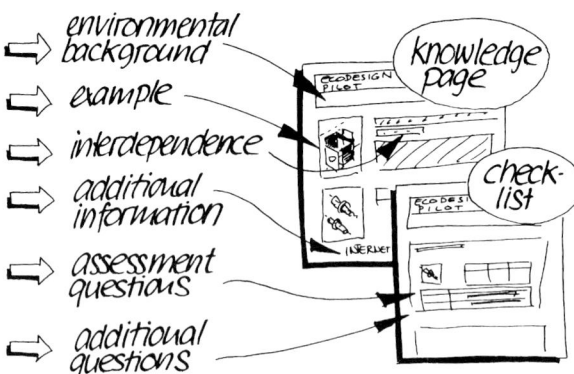

**Fig. 6 Knowledge page and checklist in the ECODESIGN PILOT (in according [1]
or the ECODESIGN online-version on: www.ecodesign.at)**

After compiling a set of successful improvement measures, the trick is in finding the necessary resources to efficiently implement these measures. The operational budget process, in other words the discussions and decisions concerning the resource distribution, is of central importance in this step (Fig. 7).

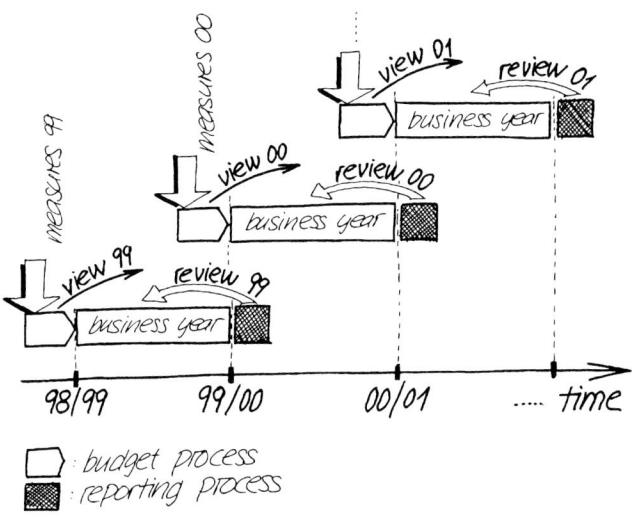

Fig. 7 Schematic illustration of "budget process" and "reporting process" in the time axis and the integration of product improvement measures (in accordance with [1], [22], [23])

The managing rhythm of an enterprise is strongly linked to the key processes, "budget process" and "process of the annual report". The chosen improvement measures are thus discussed and scheduled parallel to the other planned activities of the budget process. This ensures that the necessary re-sources can be mobilised and the improvement measures actually realised.

6 CONCLUSION AND OUTLOOK

The implementation of sustainable product design requires modified organizational structures and planning and decision-making processes as well as specific methods and resources. Life-Cycle-Thinking is of general importance which means that there is a need for adequate models and methods as well as measurement systems which support a systematic analysis of the environmental, economic, and social impact.

Presently, the authors are working on an ongoing project with the title "Delivering Research Results to the Educational Process: An Investigation of Approaches and Methodologies" within the "Alliance for Global Sustainability" (AGS)[3] . One topic is "sustainable design". Already existing educational materials are used. The idea behind is to bring together all the knowledge available at the participating schools and deliver it efficiently, using modern methods of communication and new forms of learning.

---

[3] AGS is a unique partnership with global reach involving the Massachusetts Institute of Technology (MIT), the Swiss Federal Institute of Technology in Zurich (ETHZ), Chalmers University of technology, and the Tokyo University (UT). Focused on environmental research and education, these four leading universities from three continents bring together the intellectual and technological synergy required to meet contemporary challenges. AGS works with industry, developing countries, and governments throughout the world striving toward a sustainable future. Currently, AGS has over 50 running projects. For further information, see the website at: www.globalsustainability.org

REFERENCES

1   **Wimmer, W.** and **Züst, R.** (2001) ECODESIGN PILOT Produkt-Innovations-, Lern- und Optimierungs-Tool für umweltgerechte Produktgestaltung mit deutsch/englischer CD-ROM. Zürich, Verlag Industrielle Organisation.

2   **Alting, L.** and **Legarth, J.** (1995) Life Cycle Engineering and Design. In *Annals of the CIRP Vol. 44/2/1995*. Bern, Hallwag Verlag, pp. 569 - 580.

3   **ISO 14001**: EN ISO 14001 Environmental Management Systems – Specification with guidance for use.

4   **Caduff, G.** and **Züst, R.** (1996) Increasing Environmental Performance via Integrated Enterprise Modeling. In *ECO-Performance – 3rd International Seminar on Life Cycle Engineering CIRP*. Zürich, Verlag Industrielle Organisation, pp. 39-46.

5   **Asimow, M.** (1962) Introduction to Design. In *James B. Reswick, ed. Fundamentals of Engineering Design*. Englewood Cliffs N.J., Prentice-Hall.

6   **Hall, A.D.** (1962) A Methodology for Systems Engineering. Princeton,Van Nostrand.

7   **Haberfellner, R., Nagel, P., Büchel, A.** and **von Massow, H.** (1976) Systems Engineering. 1st Edition. Zürich, Verlag Industrielle Organisation.

8   **Nagel, P., Haberfellner, R.** and **Schaper, M.** (1982) SE-Memo. Zürich: Betriebswissenschaftliches Institut (BWI) der ETHZ.

9   **ISO TR 14062**: EN ISO 14062 Design for Environment (DfE) – Guidelines for integrating environmental aspects into product development).

10 **Frei, M.** (1999) Öko-effektive Produktentwicklung. Wiesbaden, Gabler Verlag.

11 **Gerber, R.** (1997) Bestimmen der bedeutenden Umweltaspekte der Produkte der Firma Geberit. non published diploma work. Zürich, Betriebswissenschaftliches Institut (BWI) der ETH Zürich.

12 **Seipelt, D.** (1998) Bestimmen der bedeutenden Umweltaspekte von Energie- und Signalkabel der Firma Huber+Suhner. non published diploma work. Zürich, Betriebswissenschaftliches Institut (BWI) der ETH Zürich.

13 **Caduff, G.** (1998) Umweltorientierte Leistungsbeurteilung. Wiesbaden, Gabler Verlag.

14 **ISO 14031**: EN ISO 14031 Environmental Performance Evaluation – Guidelines.

15 **Dewey, J.** (1938) Logic, The Theory of Inquiry. New York, Holt.

16 **Büchel, A.** (1969) Systems Engineering. In *IO Management Zeitschrift 38(1969)9*, pp. 373 – 385.

17 **Züst, R.** (2000) Einstieg ins Systems Engineering - Systematisch denken, handeln und umsetzen. 2nd Edition. Zürich, Verlag Industrielle Organisation.

18 **BUWAL** (1996) Ökoinventare für Verpackungen. Schriftenreihe Umwelt Nr. 250, Band I und II. BUWAL (Publisher). Bern, Bundesamt für Umwelt, Wald und Landschaft (BUWAL).

19 **Ahbe, S., Braunschweig, A**. and **Müller-Wenk, R.** (1990) Methodik für Ökobilanzen auf Basis der ökologischer Optimierung. Schriftenreihe Umwelt Nr. 133. BUWAL (Publisher). Bern, Bundesamt für Umwelt, Wald und Landschaft (BUWAL).

20 **Goedkoop, M.** (1995) The Eco-Indicator 95 - Weighting Method for environmental effects that damage ecosystems or human health on a European scale. Amersfoort (NL): Pre consultants.

21 **Frei, M.** and **Züst, R.** (1998) Die öko-effektive Produktentwicklung - Die Schnittstelle zwischen Umweltmanagement und Design. In *Markt- und Kostenvorteile durch Entwicklung umweltverträglicher Produkte.* VDI-Berichte Nr. 1400, Düsseldorf, VDI Verlag, pp. 51-71.

22 **Gresch, P.** (1998) Vorlesungsunterlagen zur Vorlesung „Ausgewählte Kapitel des Umweltmanagements" an der ETHZ, Sommersemester 1998.

23 **Züst, R.** (1998) Vorlesungsunterlagen zur Vorlesung „Ausgewählte Kapitel des Umweltmanagements" an der ETHZ, Sommersemester 1998.

Sustainable prdouct development – a view from the front line

G KANE, M GELA, M G MILNER, and **G STREET**
Clean Environment Management Centre (CLEMANCE), University of Teesside, UK

ABSTRACT

Increasing legislative and financial drivers to promote sustainable development, combined with increasing levels of public awareness, have lead many companies to attempt to develop more sustainable products. This paper describes a number of sustainable product development (SPD) projects carried out by CLEMANCE with industry, mainly in Small or Medium Size Enterprises (SMEs). The paper summarises each project, the drivers behind the product involved and the barriers encountered. Themes from these examples are drawn out and discussed. The paper concludes that innovation is the key issue in SPD and that currently available eco-design tools will not deliver SPD in themselves and can even be a hindrance.

1 INTRODUCTION

The Clean Environment Management Centre (CLEMANCE) was established at the University of Teesside, part funded by the University and the European Regional Development Fund (ERDF). This initial funding provided Clean Technology support to 22 Small or Medium Size Enterprises (SMEs) between April 2000 and the end of December 2001. This paper considers a number Sustainable Product Development (SPD) projects carried out in that initial funding phase to draw out themes and lessons learnt. The paper begins by discussing various approaches to SPD before describing the actual projects carried out by CLEMANCE and the products involved. An evaluation of the products involved is carried out to determine how effective the various approaches used for SPD have been before making a number of recommendations for improvement.

2 SUSTAINABLE PRODUCT DESIGN

2.1 Overview
Charter and Chick (1) use a four level model to describe the shift from end-of-pipe to effective pollution prevention as shown in figure 1. The levels are:
- Re-PAIR: end-of pipe methods to treat an existing problem;
- Re-FINE: the use of cleaner production to produce the product for less environmental impact;
- Re-DESIGN: the use of eco-design to avoid environment problems throughout the product's life cycle;

- Re-THINK: the use of completely innovative strategies to deliver the service required by the customer.

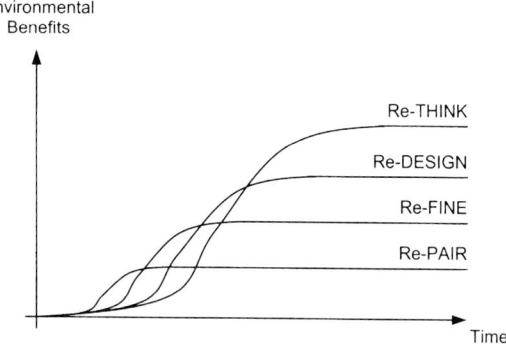

Figure 1: Strategies for Environmental Protection

The terms Sustainable Product Design (SPD) and Eco-Design are often used interchangeably, however strictly speaking SPD differs from Eco-design in two ways:
- Social issues are considered during the design process;
- The products aim to be sustainable rather than 'more sustainable'.

Charter (2) defines SPD as follows:

"SPD is a design management practice which has as a key element the need to ensure that economic, environmental, ethical and social needs are balanced. SPD acknowledges the need to develop innovative product and service concepts ... The aim is to produce zero emissions."

These rigorous requirements place SPD firmly at the Re-THINK level.

A large number of tools and methods have been developed to promote various elements of SPD. The leading examples of each are described below under three headings:
- Design Philosophy Approaches: design philosophies to develop sustainable products from first principles;
- Quantitative Approaches: those attempting to quantify the environmental performance of products to compare design options or identify particular environmental 'hotspots';
- Qualitative Approaches: 'how to' manuals describing some of the techniques to reduce the environmental/social impact of products.

2.2 Design Philosophy Approaches

2.2.1 Eco-Efficiency and Factor 'X'
Eco-efficiency is the mount of utility extracted from each unit of natural resource (also known as 'resource productivity'). A famous book, 'Factor Four' (3), advocates the use of the eco-efficiency approach to extract twice as much utility from half as much resources, thereby increasing the standard of living while reducing environmental impact. This target has been raised from four to ten by the establishment of the 'Factor Ten Club' of leading environmental thinkers. Raising the stakes even higher, Factor 20 has been mentioned in some quarters (4).

Eco-efficiency has become very popular with industry, as it can deliver significant financial benefits, however it has several limitations:

D002/004 © IMechE 2002

- It takes no account of the thermodynamic sustainability requirement for a cyclic system powered by an external energy source (see the eco-system requirements in 2.2.3);
- Eco-efficiency can improve while environmental damage gets worse, if the 'utility' produced by the system increases without offsetting other environmentally damaging processes (this is known as the 'rebound effect');
- It makes no allowance for toxicity. Some substances are highly toxic to eco-systems in very small doses.

2.2.2 The Natural Step

The Natural Step Framework (5) is used by some companies (eg IKEA) for product development. The framework stipulates four system conditions for a sustainable society:
- Substances extracted from the earth's crust must not systematically increase in nature;
- Substances produced by society must not systematically increase in nature;
- The physical basis for the productivity and the diversity of nature must not be systematically diminished;
- We must be fair and efficient in meeting basic human needs.

While the natural step approach fulfils all the requirements for environmental and social sustainability, it does not reflect the intermediate benefits to be achieved by the eco-efficiency approach.

2.2.3 Bio-Thinking

Bio-Thinking (6) combines the two previous sets of principles to derive five basic design requirements for sustainable products:
- Cyclic: The product is made from organic materials, and is recyclable or compostable, or is made from minerals that are continuously cycled in a closed loop.
- Solar: The product uses solar energy or other forms of renewable energy that are cyclic and safe, both during use and manufacture.
- Safe: The product is non-toxic in use and disposal, and its manufacture does not involve toxic releases or the disruption of ecosystems.
- Efficient: The product's efficiency in manufacture and use is improved by a factor of ten, compared to products providing equivalent utility did in 1990.
- Social: The product and its components and raw materials are manufactured under fair and just operating conditions for the workers involved and the local communities.

The first three requirements, known as 'ecosystem' requirements are those required to be environmentally sustainable. 'Efficiency' is included because of its potential environmental and business benefits and the 'Social' requirement covers the social aspect of sustainable development.

2.2.4 Life Cycle Thinking

Life Cycle Thinking is a fundamental aspect of SPD. It requires the product developer to consider environmental and social impacts across the entire product life cycle, ie 'cradle to grave'. To promote cyclic thinking, the life cycle is sometimes described as 'cradle to cradle'. It has also been proposed that for certain products (eg process plant, oil platforms) the life cycle should be extended beyond disposal to include a 'legacy phase' to cover environmental and social issues in the long term (7).

2.3 Quantitative Approaches

2.3.1 Life Cycle Assessment

Life Cycle Assessment (or Life Cycle Analysis) maps the flow of materials and energy through a product's life cycle from cradle to grave. It has been practised for many decades on an ad-hoc basis but became a structured method when the Society for Environmental Toxicology And Chemistry (SETAC) published "Guidelines for Life Cycle Assessment: A Code of Practice" (8). Since then the SETAC Guidelines have been developed and adapted for the ISO 14040 LCA Standard (9).

The advantages of the LCA approach are:

- Its holism, considering all impacts, upstream and downstream;
- The most severe impacts ('hot spots') can be identified for potential reduction;
- The results are transparent and can be easily illustrated;
- Environmental trade-offs can be made allowing 'spend to save'.

However LCAs have the following disadvantages:

- Executing even the simplest LCA takes a large amount of time and effort;
- The results are highly dependent on the definition of the 'system boundary';
- The results are highly dependent on data quality and data origin.

2.3.2 Eco-Compass

The DOW Eco-Compass (10) considers six environmental parameters and rates them from 0 to 5 (where five is the ideal) and represents the results on a 'spider-web' diagram (see figure 2). The outside boundary represents the ideal eco-profile, while the shape demonstrates how far the design is removed from the ideal. The main drawback of this approach is the emphasis on eco-efficiency measures. The eco-system requirements for sustainability are omitted.

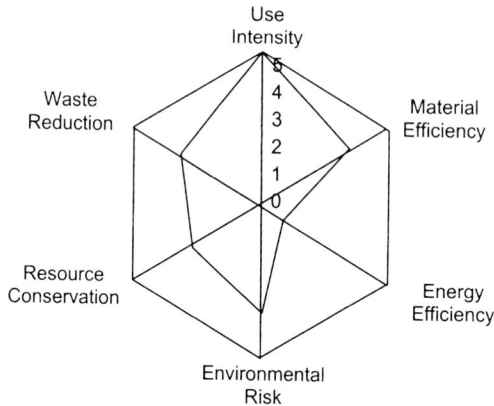

Figure 2: DOW Eco-Compass

2.3.3 Proxy Indicators

Proxy indicators use a single parameter of the design as an indication of the product's total environmental impact. While they are easy to measure and understand, they tend to promote eco-efficiency at the expense of the eco-system requirements. Examples include:

D002/004 © IMechE 2002

- Energy Consumption is often used as an environmental indicator, but usually either in conjunction with other indicators or for buildings and manufacturing plants. The assumption behind this approach is that the energy consumption is the major cause of many of the world's environmental problems (eg global warming, acidification), therefore reducing energy consumption will be the most effective way of cutting a product's environmental impact.
- Embodied Energy of a product is the total energy involved in the extraction and purification of all constituent materials and production processes involved in the manufacture of that product. This measure is used to choose materials in the eco-design of buildings (11).
- Gross Energy Requirement (GER) is the total energy consumed by a product from cradle to grave.
- The Environmental Rucksack of a product is the total mass flow over its lifecycle (3). Thus the environmental rucksack of a 12g gold ring may be 3 tonnes due to the large amount of material involved in the extraction and purification of the gold in the early stages of the supply chain. This approach promotes dematerialisation, however it fails distinguish between the toxicity of materials.
- The Material Intensity per unit Productive Service (MIPS) (12) is the total mass flow over the product's life cycle per unit utility produced by the product. In other words, it is the eco-efficiency version of environmental rucksack and, as such, is useful for Factor Four/Ten calculations
- A recent development (13) uses the thermodynamic concept of exergy as a proxy indicator. Exergy combines energy and material value into a single measure, thus providing a more 'holistic' approach than, say, energy. It is normally used to optimise the design of thermal systems, but by including environmental acceptability into the 'ground state' of a material stream, the analysis can also measure that stream's environmental impact. Exergy values of many common materials and fuels are available to carry out this type of analysis.

2.4   Qualitative Approaches

2.4.1 Design For X
Design for X (DfX) is a qualitative approach to eco-design. There are many DfX methods available, some examples are (14, 15):
- Design for Assembly: guidelines to optimise the design to minimise waste during the manufacturing process;
- Design for Disassembly: guidelines to design easy-to-disassemble products so they may be separated into easily recyclable/disposable parts;
- Design for Material Conservation: guidelines to reduce the amount of material in the product, reduce scrap and use of recyclable materials;
- Design for Upgradeability: guidelines to extend product lifecycles by making each product upgradeable to meet new tasks or performance requirements;
- Design for Re-Manufacture: guidelines to extend product life cycles by making components easily replaceable on wear out.

Many of the DfX approaches have common principles such as the use of recyclable/recycled materials or reducing the number of components. McAloone & Evans (16) argue that the 'one-function-to-one-function' of DfX is contrary to the 'whole life thinking' requirement. Instead a more holistic approach to eco-design is required.

2.4.2 Grey and Black Lists

Many companies use 'Grey' and 'Black' lists to improve environmental performance of products, for example Volvo (17, 18). Black list substances should never be used in the design and those on the grey list should only be used where no alternative exits. As a guide these lists are of benefit to an engineer, but only within a more holistic SPD/eco-design framework.

3 CASE STUDIES

3.1 Paint Stripper

Standard paint stripper has as its main constituent dichloromethane, a highly toxic substance. The alternative paint stripper involved in this project uses a combination of much less toxic chemicals. CLEMANCE carried out a review of the proposed constituents and found that one, while much less toxic than dichloromethane, is still toxic to fish. Thus it would have been very difficult to market the product as 'sustainable'. This illustrates a particular difficulty in improving the environmental performance of products; a product can be much better than the current products, but still harmful. Restrictions on labelling such as the 'Green Claims' code (19) can mean that that benefit may not be realised. Fortunately an alternative chemical has been identified and is currently being investigated.

3.2 Graffiti Remover

The environmentally friendly graffiti remover has been developed to be safe and produced from naturally occurring materials. Its unique selling point, however are its social benefits. As the product is much safer than solvent based equivalents and does not require as much personal protection equipment or training, it is marketed to probationary services so offenders can make amends for their crimes, and neighbourhood watch groups so communities can take care of their local environment. CLEMANCE's input to this product and its marketing has been very small, but it has been included as a good example of a Sustainable Product produced by an SME.

3.3 Biological Cleaner

This biological cleaner has been developed to replace a wide range of liquid cleaning fluids. Based on enzymes and surfactants, the product replaces traditional cleaners containing acids, caustics and alkalis. The cleaner continues to work 10-14 days after application, breaking down hydrocarbons and killing bacteria. CLEMANCE investigated general environmental issues relating to the product and a particular issue relating to its use in food preparation areas. The major area of uncertainty in the product's environmental performance is the environmental impact, if any, of the surfactants. This will require further investigation.
As the cleaner is less harmful to the user than standard products, it can be argued to have a positive social impact.

3.4 Power Supply

CLEMANCE has carried out a Life Cycle Assessment on an electronic power supply as a basis for its eco-design. The following barriers were encountered:

- The very large amount of time and effort required to compile information, even for a relatively small product;
- Misunderstanding of the requirements of the exercise by the supply chain;
- A perceived reluctance of the supply chain to divulge such information;

D002/004 © IMechE 2002

- A lack of publicly available generic Life Cycle Inventory data on electronic components.

The resulting uncertainty in the LCA meant that few specific conclusions could be drawn. The two environmental hotspots identified were:

- Aggregated materials production for all the components (no one component dominated);
- Energy leakage during the use phase.

Publicly available qualitative DfX checklists were also investigated for suitability. While a number of practical recommendations were made as a result of the project, it was concluded that:

- The LCA took a large amount of effort without arriving at any conclusion that could not have been predicted in advance;
- That the DfX checklists were only suitable for optimising the product, as opposed to developing radically new design concepts.

The companies involved in the project are currently considering the recommendations.

3.5 Organic Plant Food

This project involved developing a liquid plant food from chicken manure. The product specification was that it had to be cheap to produce, safe to humans and plants, and organic to Soil Association standards. The companies involved had previously proven that their method could successfully transfer nutrients into a liquid form, but CLEMANCE found unacceptable levels of phyto-toxicity and pathogens in the product. In a series of laboratory experiments, CLEMANCE developed a more effective process leading to the removal of pathogens and the toxicity. The only barrier to successful product production is the lack of a cheap organically certified carbon source, a vital ingredient to achieve a satisfactory end product.

3.6 Eco-centre

This project considered the renovation of a semi-derelict building into an eco-centre for a community-based environmental charity. The centre will provide office accommodation for the charity, classrooms for their community activities, and a display area for raising awareness for the public. CLEMANCE used a combination of literature searches, product searches and checklists to develop a series of outline design concepts. To distinguish between materials, embodied energy was used as an environmental indicator.

Due to the fact that the structure of the building was to be retained, a number of the better eco-building options could not be adopted. For example, passive solar heating systems would not be as effective as they could be in a new building.

Consideration is currently being given to demolishing the existing building and incorporating the materials into the fabric of a new purpose designed eco-centre. While this option would improve the environmental performance of the building, it may cause a negative social impact as the existing building has some historical value.

4. EVALUATION OF CASE STUDIES

4.1 Evaluation Method

In the development of these products, neither the companies nor CLEMANCE used a formal design philosophy, but rather used individual tools and information on an ad-hoc basis. In order to take a more structured approach, CLEMANCE has since carried out an exercise to identify a suitable philosophy. Of the four design philosophies considered (as described in Section 2.2), Bio-Thinking was chosen as the most appropriate due to its blending of the ecological system requirements and social justice of the Natural Step with the more business oriented eco-efficiency approach. It is also very simple to understand and apply, a particularly useful attribute when working with SMEs.

Table 1 shows the simple guide used to compare the new product to the one that it was replacing.

Table 1: Rating Definitions

Rating	Definition
✓✓✓	Impact of new product negligible in comparison to the standard product
✓✓	Substantial performance improvements over the standard product
✓	Slight performance improvement over the standard product
	Performance no better than standard product
✗	Performance worse than standard product
NK	No data or information available

4.2 Evaluation Results

By evaluating the projects against the Bio-Thinking principles as described above, the performance of the new products in terms of sustainability can be compared (see Table 2). Note that the rating for the Power Supply is that expected if the recommendations of the project were acted upon and that Eco-centre rating is for the purpose built option.

Table 2: Product Performance Summary

Product	Cyclic	Solar	Safe	Efficient	Social
Paint Stripper			✓✓✓		NK
Graffiti Remover	✓✓✓	✓✓	✓✓		✓✓
Cleaner	✓✓	✓✓	✓✓		✓
Power Supply	✓		✓✓	✓✓	NK
Plant Food	✓✓✓	✓✓	✓✓✓		NK
Eco-centre	✓✓	✓✓	✓✓	✓✓	✓✓

The following points can be drawn out:

- The popularity of the eco-system issues suggest that they are foremost in the mind of the (SPD untrained) product developer;
- The 'safe' criterion was the one most addressed by the projects. This could be due to an anthropocentric tendency either within product developers, or in their target markets;
- The 'solar' criterion was the least popular of the eco-system criteria. Several of the products consist of natural products which are produced using solar energy, but little consideration has been given to other energy use, say in manufacture and transportation;
- In fact, the 'whole life cycle' philosophy was not applied in general;
- Eco-efficiency was only considered in projects starting from an existing company product or artefact;
- Some of the projects had positive social connotations, the cleaner, the graffiti remover and the eco-centre. These were positive impacts above and beyond the definition of the social requirement in Bio-Thinking. The social impacts of the other products were unknown.

4.3 Comparison with Other Studies

Van Hemel (20) has compared her study of the SPD principles that Dutch SMEs use with those found in two other studies. In one of these, Smith investigated British, US and Australian companies of various sizes. In the other, Hanssen investigated six Nordic Companies. Van Hemel's comparison is shown in Table 3:

Table 3: Van Hemel's Summary of Use of SPD Principles

Issue	van Hemel	Smith	Hanssen
Recycling of Materials	✓	✓	✓
High Reliability and Durability	✓		
Recycled Materials	✓		✓
Lower Energy Consumption	✓	✓	
Remanufacturing	✓		
Less Production Waste	✓	✓	
Cleaner Production	✓	✓	
Reduction in Weight	✓		✓
Cleaner Materials	✓	✓	✓
Packaging Issues	✓		

The two issues that appear in all three studies are 'recycling of materials' and 'cleaner materials' which relate to the 'cyclic' and 'safe' criteria of Bio-Thinking. This emphasis reflects the evaluation carried in this paper.

5. DISCUSSION

None of the companies involved in these projects made any use of the approaches and tools described in Section 2 unless they were introduced by CLEMANCE. However, the products scored well in the Bio-Thinking evaluation above, supporting the idea that the key issue in Sustainable Product Development is innovation. Therefore a clear design philosophy is required from the outset to ensure that the innovation addresses all SPD issues. CLEMANCE has found the Bio-Thinking approach gives the optimum coverage of sustainability issues in an easy to understand format.

Many commentators, eg Mann (21), feel that groundbreaking, radical innovation tends to occur in SMEs (eg Dyson, Hotmail), as large organisations (eg Hoover, Microsoft) are less adaptable and tend to focus on optimising products or buying (or even 'borrowing') new design concepts from SMEs. This is contrary to the commonly held view that, due to their financial muscle, large companies have a monopoly on technical innovation. In CLEMANCE's experience, this resource advantage is often outweighed by the 'institutional inertia' inherent in many large organisations. It is notable that many of the SMEs involved in these CLEMANCE projects were established specifically to develop and produce one particular sustainable product.

Eco-design tools such as LCA and DfX checklists are of little use during the generation of new design concepts, in fact it can be argued that they can stifle creativity by focussing too much on the existing product (eg LCA) or complexity (checklists). The place for these tools is later in the design process, for example, LCA can be used as an evaluation tool once concepts have been developed or to identify environmental hotspots for optimisation. Checklists can be used as a source of best practice when developing the detail of each design.

Considering the case studies against the Bio-Thinking criteria, the foremost concerns of the product developers were the three eco-system criteria: 'solar', 'cyclic', and 'safe', with a particular emphasis on the latter. This is encouraging as these criteria are regarded as essential for sustainability. Life Cycle Assessment mainly considers the 'safe' and 'efficient' criteria. To a certain extent it takes 'cyclic' into consideration by offsetting future material extraction against any material in the product expected to be recycled. However it is felt that the method does not adequately 'reward' advances in 'cyclic' and 'solar' aspects of the product.

With the exception of the LCA project, the 'life cycle thinking' requirement is not adequately addressed in the projects studied.

6. CONCLUSIONS

The conclusions of this paper can be summarised as follows:
- Innovation is the key issue in Sustainable Product Design;
- Conventional tools cannot stimulate innovation and can actually hinder it;
- Having the correct design philosophy is much more important than use of these tools;
- The sustainable product revolution will probably take place in SMEs as that is where radical innovation can take place;
- In the experience of CLEMANCE, the Bio-Thinking approach is particularly suitable for SMEs and will be adopted in future projects;
- The 'solar' element of Bio-Thinking needs to be given more attention;
- The 'life cycle thinking' element of SPD philosophies needs to be given a higher priority.

CLEMANCE is currently addressing these issues by developing a SPD resource pack and an eco-innovation methodology for SMEs. When complete, the resource pack will be freely available from the CLEMANCE website.

REFERENCES

1 Charter, M., Chick, A. (1997) Editorial, Journal of Sustainable Product Design, No 1, April 1997, pp 5-6
2 Charter, M. (1997) Managing Eco-design, UNEP Industry and Environment Vol 20, pt 1-2, pp29-31
3 von Weizsäcker, E., Lovins, A.B., Lovins, L.H. (1997) Factor Four: Doubling Wealth, Halving Resource Use, Earthscan, London, ISBN 1 85383 407 6
4 van Weenan H, 1999: Design for Sustainable Development, European Foundation
5 The Natural Step (2000) What is the Natural Step, internet publication: http://www.naturalstep.org.uk/whatis.thm
6 Datchefshki, E. (1999) Cyclic, Solar, Safe – BioDesign's Solution Requirements for Sustainability, Journal of Sustainable Product Design, Issue 8, January 1999
7 Kane, G., Stoyell, J.L., Howarth, C.R., Norman, P.W., Vaughan, R. (2000) "A Stepwise Life Cycle Engineering Methodology for the Clean Design of Large Made to Order Products", Journal of Engineering Design, Vol 11, No 2, pp175-189
8 Consoli, F., Allen, D., Boustead, I., Fava, et al (1993) Guidelines for Life-Cycle Assessment: A Code of Practice, SETAC, Brussels, ISBN 90-5067-003-7
9 ISO 14040: Environmental Management Systems - Life Cycle Assessment - Principles and Framework, British Standards Institute
10 ENDS 252 (1996) Dow Europe and the Challenge of Eco-Efficiency, ENDS Report 252, January 1996, pp16-19
11 AECB (2000) The Real Green Building Book 2000, Association of Environmental Conscious Building (AECB)
12 Schmidt-Bleek, F. (1993) MIPs-A Measure of Progress towards Sustainability, The LCA Sourcebook, Society for the Promotion Of LCA Development
13 Finnveden, G. (1998) Characterisation of Resources by Exergy Consumption or Entropy Production, 8th Annual Meeting of SETAC, Bordeaux 14-18th April 1998

14 Fiksel, J. (1996) Design for environment: creating eco-efficient products and processes, New York: McGraw-Hill, 1996, ISBN 0070209723

15 Mackenzie, D. (1997) Green Design: Design for the Environment, Laurence King Publishing, London, ISBN 1 85669 096 2

16 McAloone T, Evans S, 1995: The Challenges of Environmentally Conscious Design International Conference on Clean Electronics Products and Technology, (CONCEPT), Institute of Electrical and Electronic Engineers, October 1995

17 Volvo (1998) Chemical substances which must not be used in the Volvo group: Volvo's black list, Internal report STD 1009,1

18 Volvo (1998b) Chemical substances whose use within the Volvo group shall be limited: Volvo's grey list, Internal report STD 1009,11

19 ISO 14021: Environmental Labels and Declarations - Self-declared Environmental Claims, British Standards Institute

20 van Hemel, C.G., (2001) What Sustainable Solutions Do Small and Medium Size Enterprises Prefer?, Sustainable Solutions, Greenleaf Publishing 2001

21 Mann, D., (2000) "The Four Pillars of TRIZ", Engineering Design Conference 2000, Brunel University

Implications of the integrated product policy (IPP) in new products design and development

M SORLI
Mechanic Unit, Labein Research Center, Bilbao, Spain
R ZUBIAGA
Environmental Unit, Labein Research Center, Bilbao, Spain
J A GUTIÉRREZ
Enterprises Organisation, University of Basque Country, Bilbao, Spain

ABSTRACT

Unavoidable trend toward developing more eco-friendly products within a sustainable growth scenario, handles several implications from all entities related to the product life-cycle: product value-chain actors, consumer/user and most likely, new services for Retrieving, Re-using and Recycling, practically non-existing nowadays.

Recently issued Green Book on Integrated Product Policy (IPP) by the European Commission (7/02/2001) basically propose several parallel action lines:

- Creating an environmental culture among all actors along the product life-cycle: manufacturer, users (by eco-utilization), after sales service on his more comprehensive meaning (from the manufacturing gate to the product end-of-life; instead of the old: *From cradle to grave*, this should be: "*From leaving-home to grave*")
- Fostering programs from the Government bodies and general administration aiming to incentive manufacturers to designing and producing eco-friendly products, as well as incentiving consumers to purchasing green products, etc.
- Fostering local administrations to create, promote and subsidize the necessary end-of-life (Retrieving, Re-using and Recycling) systems.
- On the contrary, creating new mechanisms to penalise excessive environmental costs based on the principle: "Polluting has to pay".

Confining the problem just to the boundaries of industrial design which is the most important phase on the product life (*birth*) where actual performance of the product in *real life* is fully committed several questions arise as: What are the implications of this unavoidable approach from the perspective of industrial development and design units? How can existing tools and methodologies help industrial designers to cope with all these new rules and standards to come?

Present paper pretends to give an overview of this new scenario and to present a theoretical proposal of the product development process including all these new issues.

1. INTRODUCTION

Lead Time or *Time to market* has been generally admitted to be one of the most important keys for success in manufacturing companies. Combination of factors such as ever changing market needs and expectations, rough competition and emerging technologies among others, challenges industrial companies to continuously increase the rate of new products to the market to fulfil all these requirements.

Besides that, current socio-economic trends towards developing more eco-friendly products within a sustainable growth scenario, add a new complexity to the manufacturing companies mostly but not only, to the product designers.

2. INTEGRATED PRODUCT POLICY (IPP)

Recently issued by the European Commission (7/02/2001), the Green Book on Integrated Product Policy (IPP) (1) basically proposes creating an Europe-wide strategy in order to reinforce and redirect product oriented environment policies. This strategy bases on the Integrated approach (IPP) and aims to complement other environment policies in order to improve products and services along their life-cycle by means of exploiting their unveiled potential.

The policy should base on several parallel action lines:
- Creating an environmental culture among all actors along the product life-cycle: manufacturer, users (by eco-utilization), after sales service on his more comprehensive meaning including disposal at the end-of-life.
- Fostering programs from the Government bodies and general administration aiming to incentive manufacturers to designing and producing eco-friendly products, as well as incentiving consumers to purchasing green products, etc.
- Fostering local administrations to create, promote and subsidize the necessary end-of-life (Retrieving, Re-using and Recycling) systems.
- On the contrary, creating new mechanisms to penalise excessive environmental costs based on the principle: "Polluting has to pay".

On the authors' opinion, these proposals however being full consequent and absolutely necessary, need a wide time range for their implementation. Doing a summary analysis of the Green Book suggestions, there are several aspects to be developed prior to achieving real results:

- Creating the cultural aspects implies not only education, training and real commitment from all actors along product life-cycle but also (quite crucial) providing means allowing these actors to fulfill eco-requirements in a convenient low cost way. Maybe the most important point on this issue is the creation of necessary services for recycling, disposal, etc. These facilities will not arise from the private initiative until they will actually become real making money businesses. It seems that administrations will have to support and subsidize their early phases so enabling the process to start up.

- Regarding the other aspects both in positive (to subsidize end of life systems) and in negative (to penalize high environmental costs), they are completely in the hands of politicians from the different Governments of the Member States and the local administrations. In summary they will require accomplishing several steps (consensus among political parties, generating legislation, allocating resources, etc.) before reaching the operative level.

It appears finally that by now, the process of new products development and design is the only aspect on which immediate actions may be undertaken in order to improve environmental performance of new products.

3. PRODUCT DESIGN AND DEVELOPMENT

Industrial design is generally admitted to be the most important phase on the product life (*birth*). Every aspect of the future performance of the product is fully committed in this phase. Decisions in early phases are easy to be taken but future consequences of wrong ones are very difficult to foresee unless a comprehensive and coherent design process had been set up.

Eco-decisions and environmental impact has for sure to be integrated into this bunch of vital decisions. Within this approach, the above mentioned Green Book on IPP should soon became the basic *Bible* for industrial product designers.

The objective of any design process should be **to concentrate changes in the preliminary Design stages**, in such a way that the product is mature by the time it reaches manufacturing stage – in which changes bring heavy costs – and reaches the market in virtually **"untouchable"** form. Furthermore, any change when the product is already in the market is regarded as dramatic on account of their brand-image repercussions.

The next graph (Figure 1) gives us a very significant picture of the comparative levels of **Real Costs** incurred throughout the development process and the **Compromised Costs** – i.e. costs that depend on decisions taken during the process and that "mark" the life of the product. Thus we see that the real costs in the early stages of design are modest, but decisions taken then actually "compromise" the future cost of the product/process throughout its service life.

According to the Integrated Product Policy approach, it is evident that "life-cycle costs" (not very much cared about until now) are becoming in the near future the most relevant cost item.

The Model that we are going to describe now, cover all mentioned issues and provides a comprehensive framework for a consistent new product development and design facing the challenges of the XXI Century.

DESIGN PROCESS

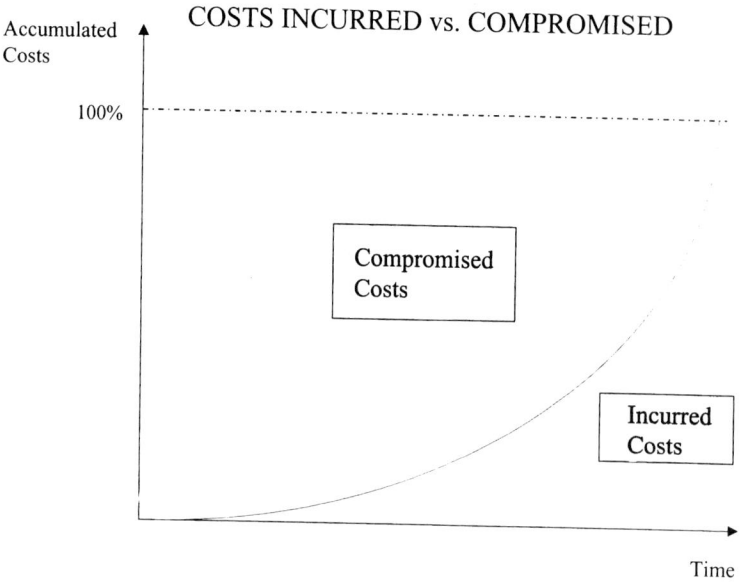

Figure 1   Design Costs

4 GENERAL STAGES IN THE NEW PRODUCT-DESIGN MODEL (2)

4.1   General introduction

The Model we are putting forward brings together techniques drawn from:
- Concurrent engineering
- Total quality management
- Extended enterprise (suppliers and users integration)
- Information and Communication Technologies (ICTs).

The **Model** puts forward a framework comprising methodology, guides for action, and tools based on the integration of above mentioned factors, all underpinned by an information-management system. This Model is to serve as an aid in developing parts in a complex, heterogeneous context featuring the participation of a number of companies in the supply chain in addition to the various internal sections of the company. New environmental issues will include even those on charge of the disposal of the product).

Furthermore, the concept of eco-friendly products from the IPP approach brings on another very important issue to be integrated in the new product development: Design has to consider and evaluate every possible environmental implication on the product/process life-cycle.

                                   D002/008 © IMechE 2002

Under this idea the **Model** proposed here consists of getting *all the departments* involved in product design and development to work closely together from the initial stages bearing constantly in mind the objectives of optimising the added value (including in this concept the environmental concerns) and reducing the number of changes. Within this new approach, *all departments* means internal and out of the company teams: "**Extended Enterprise**".

In short, it can be said that the main characteristics of this new design and development process are based on three fundamental aspects on superimposed levels, namely:

Gearing company culture towards Total Quality (3) & (4)

This entails a number of changes which, to summarise from several sources, are framed in following *ten commandments:*

1. Customer orientation.
2. Management leadership.
3. Decisions based on analysing the facts and the data.
4. Management by processes.
5. Involvement of all the staff (from the "Extended Enterprise")
6. Quality assurance for the product.
7. Association with suppliers. ("Extended Enterprise")
8. Looking for results not only in the short, but also in the medium and long term.
9. Continuous improvement.
10. Contributions to society.

Changes in operations and in organisation structure

Teamwork

Work teams showing the usual features of present-day teams – **interdepartmental and multifunctional** – plus adding the new feature of also being **inter-company**: "extended Enterprise".

Thus the work teams are made up of staff who, in organic terms, belong to different departments of the leader company *(interdepartmental)* and of the cooperating companies *(inter-company)*, and who also contribute their knowledge and experience in different disciplines *(multifunctional)*.

Concurrent Engineering

Concurrent Engineering is based on making the process stages overlap and concur, using the team structure mentioned above in conjunction with advanced tools and methodologies: QFD, TRIZ, DoE, FMEA, Taguchi, 7 M´s, Simulation, etc., all backed by an Information and Communication Management System (ICT).

Information Management

Although the structure of the teams itself together with their increased autonomy, usually facilitates communication, it is very important to establish the appropriate communication channels very clearly. It is also very important to be fully aware of usual human problems (incompatibilities, varying attitudes towards work and cooperation, personal conflicts etc.) in order to prevent them (e.g. at the team-forming stage) or to face and deal with them quickly, when they arise. Evidently, this is one of the main duties of the Management – both the **formal** heads (hierarchical structure) and the **operational** heads (team leaders).

The new Information Management System requires certain fundamental elements that will be discussed in further detail in point 5.

4.2   Defining the Specifications

In this stage, the characteristics of the new desired product and the way how it is to be manufactured are mapped out. Consequently, eco-features are a very important set of these characteristics.

GATHERING REQUIREMENTS FROM:

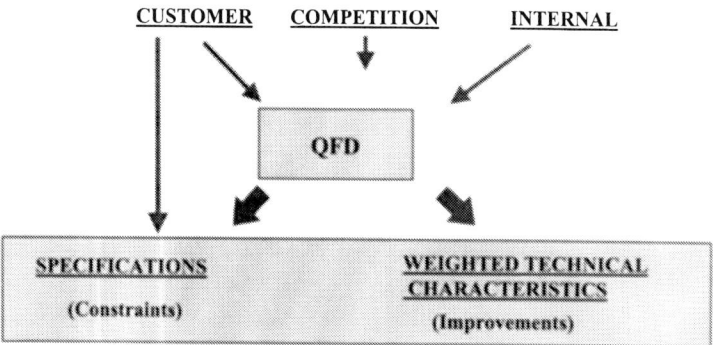

Figure 2   Defining the Specifications

The key tool at this stage (see diagram 2) is QFD (see point 6.1 later) which helps us to interpret and organise the various groups of requirements, the output being the set of product specifications. The inputs for the QFD process come from three main sources: the customer (the most important source); internal needs or policies; and the situation of the competition.

A very relevant group of needs is clearly related to the environmental issues. These needs may arouse from the customer itself (given the increasing people awareness on environmental issues), from policies set up by the governments and local authorities and even from the company's own policies.

                                                                                          D002/008 © IMechE 2002

The specifications termed 'basic' or 'restrictions' are mainly directly imposed by the customer, who sets down a number of minimum requirements in his tender specifications. The "would-be improved" specifications are those that enable us to identify the main points which, if improved, will bring a significant increase in satisfaction in the market. Naturally, "over-exciting" specifications drive that satisfaction level up exponentially.

Though environmental aspects are nowadays more close to the "over-exciting" specifications (something not expected in the product but giving great satisfaction if found), they are shifting very rapidly to the "would-be improved" and will likely become "basic" in very short term.

4.3. Conceptual Design

By **Conceptual Design** we mean designing a product in general terms, covering aspects such as the architecture and modularity of the system, its main and secondary functions, volumes, interfaces with other elements around it etc. One or more possible conceptual solutions will be analysed and selected in this Conceptual Design, for subsequent validation.

However, it is highly advisable to define as little as possible at this stage as regards technological and functional solutions, materials to be used, processes etc., since if such aspects are demarcated too tightly here, the range of possibilities become limited and creativity is blocked. As a result we get routine, repetitive solutions – in a word, bad solutions.

Starting from the set of specifications, two distinct paths can be traced for reaching the Conceptual Design of the product: Product Improvement (Redesign) or New Design. We take New Design to be the launch of a radically new product, so new that it may even entail a change in the line of business – making use of the organisation's core competencies to change the product while staying with the same technology, or introducing a sizeable number of changes in the current product, ushering in new product generations.

The distinction between Redesign and New Design is difficult to pin down generically, though designers clearly make the distinction for a specific product. In the automotive world, a fairly clear distinction is made between "restyling" (involving a number of 'cosmetic' changes made to the current model without changing its overall conception, and keeping the same name for the model), and 'New Model' (usually involving a change of product name). In general, we can say that the level of changes introduced is what makes the difference. As said before, the real point is customer perception: *if the customer feels the product as really new, then it is new.*

We thus view **"Product Improvement"** (Redesign) as developing a "new product" based on the current one through introducing a number of relatively small improvements whereas **"New Design"** implies developing and designing a significant new product with a clear and tangible difference from the old one.

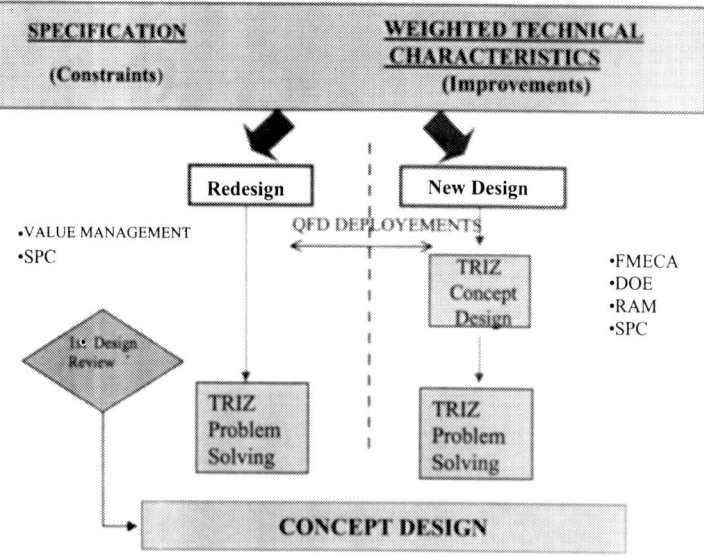

Figure 3   Conceptual Design

4.4   Detail Design

A variable and interactive set of stages runs between Conceptual Design and Detail Design, we having broken these down into:

- **Simulation**: The use of tools, essentially computer tools, to "fill in" the Conceptual Design with technology and constructive solutions. Simulation and computation are used to validate the conceptual solutions, the most promising one being chosen.
- **Improving the solution:** using conceptual tools to improve and optimise the chosen solution as far as possible.
- **Prototyping:** developing real or virtual prototypes for evaluating the product's performance in the real world as realistically as possible: structural representation, fatigue tests and trials, strength, matters of appearance, interferences with the surrounding context, ergonomics etc.

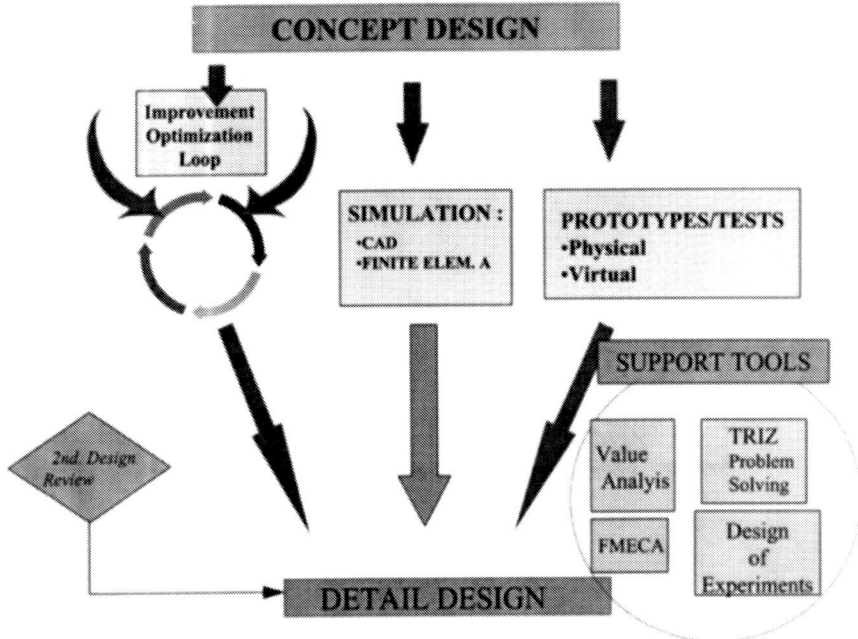

Figure 4   Detail Design

As we have said, this process may go through all the interactions deemed necessary until a suitable solution is reached, within the time limits set by the overall framework of the development process.

5 INFORMATION AND COMMUNICATION TECHNOLOGIES (ICT)

It is obvious that only last years' rapid evolution of the Technologies of Information and Communications management (ICT) has actually allowed this new working paradigm.

Big automotive manufacturers are since some time ago launching many experiments on "Virtual product development" by means of using new Internet based technologies in connecting remote locations in cooperative work on the same project.

Some aspects of the ICT tools relevant to our **Model** are:

- **The use of systems** of the kind known as *shareware* and *groupware*; drawing on a common database, these facilitate a number of aspects such as sharing information, working simultaneously with the same data, off-site working etc.

- **Management by projects:** As previously pointed out, the new decentralised approach entails a significant loss of control. The new scheme implies a change in the control paradigm. Control must be focused exclusively on the Project, on what we could call "Project Quality", and so a range of indicators must be established to enable real-time evaluation of performance to be achieved: progress, efficiency, delays, deviations, critical points etc. This is what is known as Management by Projects. Of great interest in this aspect are the contributions of Goldratt in his books "The Goal" and "Critical Chain" (5).
- **The existence of interfaces** for successful communication – interfaces that are not limited solely to computer tools, but rather are based, to a large extent and more importantly, on human relations.

Another important benefit of ICT for multinational companies, based not only in different countries but even continents, comes from the possibility of working on a 24 hours shift. Passing the workpackages over through the *"Wide Word Web (www)"* and using common workspaces and databases, allows remote teams to jump from continent to continent linking the project on a continuos mode, from USA to India, Europe, Japan, etc. In this way design engineers may literally work around the clock. These "e-business" strategies also contribute to lead-time reduction.

Some examples of this new ways of working can be collected from Honda claiming that his all new 2001 Honda Civic model, first model to use the new strategy, has achieved a process reduction of 15%. Also Daimler-Chrysler on a supply-chain pilot claims a reduction of 92% coming down from previous 14 days to just one, on the specific issue of sending production program information to suppliers (6).

ICT give design people a sound base to build on it. Nevertheless, it is not to forget that tools of this kind are *just tools*. In summary, they need people working on them using methodologies and other conceptual tools of the kind that have been mentioned in the previous points and are described in the following ones.

6 METHODOLOGIES AND TOOLS

Along previous point's description of the **Model**, many methodologies and tools may be seen in the graphs. These are some of the most useful and powerful on the authors' opinion. Many of them are well know and widely used but there are a couple of them deserving a quick look: QFD & TRIZ.

6.1   QFD: Quality Function Deployment. (7),(8) & (9)

QFD could be defined in an academic way as "A process structured to capture the voice of the customer and convert it into product requirements in each design stage with the participation of all functions of the company intervening in the design".

In brief, we are "forced" to strictly follow a series of systematized stages, in which all functions will participate. By this way, the following goals are reached:

- Nothing is "missed". The system tries to foresee all product requirements for its entire life.

- All information is channelled from the customer ("*The voice of the customer*") towards the customer ("*The customer is the King*"). Customer needs and requirements drive internal decisions.
- The creativeness of team members is enhanced. Several techniques are used to try to "invent" new performances of the product giving greater satisfaction to the customer and allowing the manufacturer to achieve a higher competitiveness level.
- A series of tools (7 M's, FMECA, Design of Experiments, RAM Techniques, etc.) which were scarcely used until now, are applied in a joint and systematized manner.

QFD as guiding thread for the Model.
QFD has to be used as the guiding line for the whole process in the new Model. The Macroflow (figure 5) and the Deployment lines (figure 6) constitute the basement on which the new product has to be build by means of the described Model.

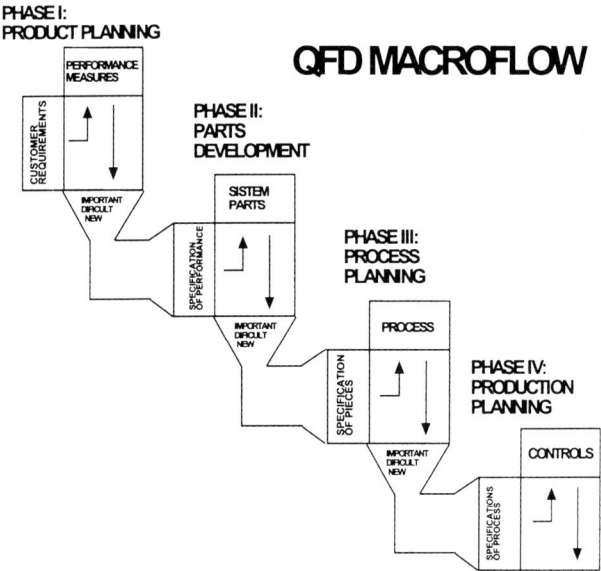

Figure 5   QFD Macro Flow

QFD Macro Flow collects and analyses customer' requirements (Phase I) and keep them alive along the whole process to the production phase in terms of Quality Control Planning (Phase IV). Phase II transfers overall product requirements to its parts, components and assemblies while Phase III cares on the design and specification of the means and resources required for production.

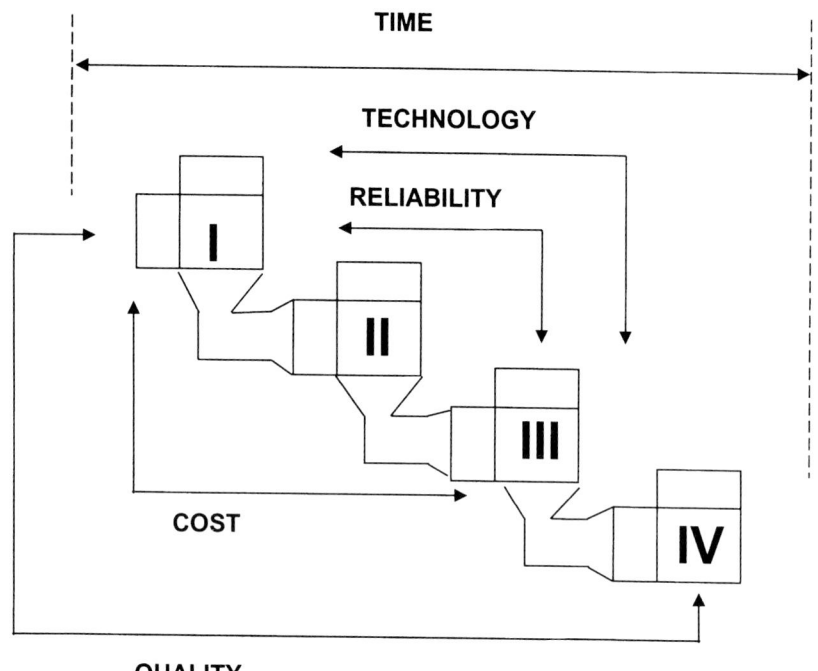

Figure 6   QFD Deployments

Macro Flow vertically transfers Customer requirements to the bottom line on the factory, creating a consistent link between any decision on the shopfloor with a related customer's need. QFD Deployments (figure 6) on the other side work on a horizontal scope, covering the whole product definition, including environmental issues, on the four main aspects of:

- **Quality**: What the customer expects and how will the manufacturer fulfil it ? Environmental issues has to be considered as one of the most important aspects of quality: "**eco-quality**".
- **Reliability**: How is the product going to perform along time and which is its life expectancy ? Surely enough, environmental features have to be kept along time in the pre-defined range.
- **Cost**: Is all this going to be achieved within the cost objectives ? Eco-design considers "cost" of the product/process as the overall cost for the humanity in terms of resources consumption and impact on the environment.
- **Technology**: Available technology is allowing us to do all that, or should we try something new ? Different technologies have different impact on environmental aspects.

Answers to these questions have to be discussed by the designer's team in an iterative mode within time constraints limiting the process. If some of the answers is negative, new assumptions have to be done, some objectives have to be downsized, or…innovative solutions have to de developed and applied. **TRIZ** (next paragraph) is a valuable tool for that.

                                                                D002/008 © IMechE 2002

6.2   TRIZ: Theory of Inventive Problem Solving (10).

TRIZ is the acronym for Russian words for **Theory of Inventive Problem Solving**, and in the current world where companies wrap up their products with more intellect than material, TRIZ is on the verge of becoming a compulsory standard.

What technology does a company use to generate breakthrough concepts and ideas? How does a company evaluate these concepts, establish priorities, and develop its strategy of innovation? Conventional tools for product improvement include Flow Diagrams, Cause-and-Effect Diagrams, Check Sheets, etc. Many companies have also included in their arsenal higher level methods as Value Engineering, Function Analysis, Robust Design, QFD, etc. Problem solving technology in all of these tools and methods is based mainly on **brainstorming** which is a good method for generating creative ideas in a team, but it depends basically on the ability of every individual for generating ideas.

The uniqueness of TRIZ's approach, compared with other methods based on psychology and management theory (such as brainstorming), is that modern TRIZ is founded on over fifty years of R&D in areas of engineering sciences and technology. Along that time, over two million patents were studied from all across the globe and in all known domains of engineering. The goal was to identify and refine recurrent inventive principles and to formulate them as a series of rules and structures transmissible across different domains.

How TRIZ Works:

Humans solve problems by analogical thinking. That is, we try to relate the problem confronting it to some standard class of problems (analogous) with which we are familiar, and for which there exists a solution. If we can draw the right analogy, we can arrive at the right solution.

Altshuller (creator of TRIZ methodology) (11) found out that the same fundamental problem (contradiction) had been addressed by a number of inventions-in different areas of technology. He also observed that the same fundamental solutions were used over and over again, often separated by many years. He sought to extract, compile and organize such knowledge. Consequently, TRIZ compiles the human innovative experience and provides access to the most effective solutions, independent of the specific technological area or industry in which the solution was developed. To do this the **principle of abstraction** is used. For virtually any inventive problem, a general (standard) problem/solution pair can be developed which does not include industry-specific information.

TRIZ helps transform this complex process into a systematic, linear process of problem solving which can be followed in a step-by-step manner. The Algorithm for Inventive Problem Solving (ARIZ) is the main TRIZ problem solving method. ARIZ divides the process of contradiction elimination into a set of steps that guides the engineer towards the ideal solution: the one closest to the system ideality.

Methodology Premises:

Basic idea behind TRIZ rests upon a small group of premises simple to understand and somewhat more complicated to apply.

- **Ideality**:

The ideal system performs a required function without actually existing. The function is often performed using only existing resources.

Table 1   Ideality

$$IDEALITY = \frac{AllUsefulFunctions}{AllHarmfulFunctions}$$

IDEALITY is a theoretical ratio among Useful and Harmful functions. Useful functions provide value to the user; harmful functions are undesired and to be eliminated whenever possible.

An important pattern discovered by Altshuller states that all technological systems evolve toward increasing system ideality. Its is not a quantitative number, but rather a qualitative assessment.

- **Contradictions:**

Technological problems result from **engineering contradictions** (improvement in one part of the system causes degradation in another). A contradiction is a situation whereby attempts to improve one feature of a system result in the degrading of another feature. Engineering contradictions result from **physical contradictions** (opposing requirements for the same characteristic). In many cases, contradictions are not immediately obvious.

There are a small number of fundamental physical contradictions. They apply across all technologies: automotive, electronic, aerospace, agriculture, etc. Most physical contradictions have been solved, without trade-off or compromise, in one technological field or another. These so-called "inventive" solutions (12) can be applied, generically, to other technological fields.

Traditionally, technical contradictions are resolved by trade-off or compromise, balancing the confronted technical parameters. TRIZ suggests just eliminating the contradiction to innovate without the use of trade-offs.

- **System approach:**

Typically, when a problem arises within a system, the engineer trying to solve the problem focuses upon the system. An experienced inventor thinks differently. He simultaneously considers the supersystem and all the associated subsystems as well. In addition, he considers each of these as they existed in the past, as well as how they might exist in the future. This approach opens a wide range of ideas that would not be easily found by traditional means. This is clearly a very complex process, and only experienced inventors are able to deal with this complexity.

  D002/008 © IMechE 2002

7. CURRENT TRENDS

As a sample of what is starting to happen with this new approach on environmental issues on the world of automotion, following comments have been extracted from "Futuring the next Industrial Revolution"(13). According to his presentation, environmental aspects of the next industrial revolution involves a transition from making vehicles "less harmful" to the environment, to making vehicles "positive" for the environment.

Nowadays, automotive industry is on the dawn of the "eco-times". For example: they are still using expensive devices like catalytic converters to minimize emissions while some emissions still represent huge potential menaces (global warming); recycling efforts are not efficient enough yet (as few as 1.045 pounds out of every 10.000 pounds of raw material used to manufacture a vehicle is currently reused). Today's materials are essentially chosen without considering the total life-cycle cost associated with disposing vehicles at end-of-life (where materials like PVC must be treated as toxic waste).

Bill McDonough and Michael Braungart (14) of McDonough Braungart Design Chemistry (MBDC) propose a paradigm shift associated with making vehicles "positive for the environment". This paradigm focuses on activities that are regenerative (versus less-bad) using principles of "Eco-Effective Design". The goals of "Eco-Effective Design" are: to meet and exceed established quality, economic, and technical performance criteria, while fostering healthy and prosperous conditions for humans and ecological systems by reutilizing materials/components in natural biological or technical cycles. Biological cycles refer to the biological metabolism of nature that reutilizes waste products such as organic matter as biological nutrients for another life cycle. Technical cycles refer to the reuse of components and other technical waste products (such as polymers, metals, organic solvents, etc.) as nutrients for another generation of products. "Eco-Effective Design" involves designing products to fit these two cycles to recapture value over and over.

The better example of "Eco-Effective Design" comes from how American Indians utilized buffalo. Literally, nothing in a buffalo was waste, every part of the animal was useful: Skin was used for making clothes, shelter, and rugs; meat for food; bones were used for decorations and tools. Even dried dung was used as fuel for fires. Every use promoted sustainability of the system in a healthful, renewable way. Similarly, every part of a vehicle must be designed to add value at end-of-life. From an Ideality point of view, there are huge opportunities associated with reducing cost, minimizing waste, and reducing energy associated with "Eco-Effective Design".

MBDC has developed a strategy and methodology for "Eco-Effective Design" of any product:
- Appropriate Materials Selection (especially eliminate materials and substances in the product and process that pose a risk to human health),
- Designing for Recycling (DfR): Design some components for refurbishment and reuse, design others to use homogeneous materials capable of high-value recycling.
- Designing for Disassembly (DfD): Simplified modular design using easy-to-release disassembly methods.

8 CONCLUSIONS

Increased use of recycled materials, European "End-of-Life" legislation associated with increased use of recycled materials and decreased use of hazardous ones, ISO 14000 and samples from companies like MBDC, indicate that the Environment paradigm shift has already begun. Our choice is between system decline and sustainable growth: "If the present growth trends in world population, industrialization, pollution, food production, and resource depletion continued unchanged, the limits to growth on this planet will be reached sometime within the next 100 years. The most probable result will be a sudden and uncontrollable decline in both population and industrial capacity" (15).

Authors believe that the new **Model** for Product Design & Development Process that have been outlined along this paper offers the industrial designer a comprehensive methodology integrating the latest technologies and tools. The **Model** will help him to deal with the increased ecological requirements covering the product (and his process) life-cycle.

There are some innovations being put forward in this New **Model** that make it unique. These innovations should be highlighted as follows:

- Integrating a novel Innovation Methodology – **TRIZ** (described above) – into the Product Development Process.
- Employing QFD (Quality Function Deployment) as the guiding thread for the entire process.
- Viewing **Integrated Engineering** as the integration of Concurrent/Simultaneous Engineering (CSE) and Total Quality Management (TQM) at Management level; these two methodologies will work together to give better quality (meeting customer requirements) and shorter development lead time.
- Developing a methodology, guidelines and tools within a multifunctional and multi-company context: "**Extended Enterprise**".
- Making use of the potential and ability of the new ICTs which allows quick and reliable exchange of information working in parallel frequently at remote locations.

REFERENCES

1. **European Commission** "Integrated Product Policy (IPP)" February 2001.
2. **Sorli, M.;** "Integration of TQM and CE techniques in the Development of New products in a multi-company framework" Ph.D. Thesis, University of the Basque Country (UPV/EHU). Spain, July 2000.
3. **Akao, Yoji**. "Hoshin Kanri. Policy Deployment for Successful TQM". Productivity Press. 1991.
4. **Merli, G.** "Total Quality as Business Tool. Strategic response to the European challenge" 1995.
5. **Goldratt, Eliyahu M**. ;"The Goal" 1996; "Critical Chain" 1997
6. **Gehm, Ryan**; "e-Business: the New Game in Town". Automotive Engineering International Magazine. April 2001
7. **Akao, Yoji;** "Quality Function Deployment. Integrating Customer Requirements into Product Design". Productivity Press. 1990.

8. **Sorli, M** and **Ruiz J**. "QFD. A Tool for the Future". Labein 1994
9. **Sorli, M**. and **Mañà, Francesc;** "Strategic Product Innovation using QFD and VM". 3rd International Symposium on QFD. Linköping (Sweden). October 1997.
10. **Sorli, M**. "Triz. A Powerful Tool for Strategic Innovation" ISPM International Conference. S.Sebastián (Spain). 1997.
11. **Altshuller, G**. "And Suddenly the Inventor Appeared". Technical Innovation Center. Inc. 1996.
12. **Altshuller, G**. "40 Principles. TRIZ Keys to Technical Innovation" Technical Innovation Center, Inc. 1997.
13. **Zlotin, Boris; Zusman, Alla**; and **Smith, Larry R**; Ford Motor Company. "Futuring the next Industrial Revolution" Workshop on Automotion. Bilbao (Spain) 2000
14. **Braungart, Michael** and **McDonough, Bill**; "The NEXT Industrial Revolution," Atlantic Monthly magazine, Vol. 282, No. 4, October, 1998.
15. **Meadows, Donella H., Meadows, Dennis L., Randers, Jorgen, Behrens III,** and **William W.**; "Limits to Growth", 1972.

CIAM and North-eastern industry – the road to sustainability

T D SHORT, J A GARSIDE, I D CARPENTER, and **E A APPLETON**
Centre for Industrial Automation and Manufacture, University of Durham, UK

ABSTRACT

With the decline in UK manufacturing, industries in the North-East of England have suffered tremendously. Many companies have been unable to meet the needs of the present, let alone consider those of the future. Members of the Centre for Industrial Automation and Manufacture (CIAM) at the University of Durham have been investigating and developing a number of tools that can help to ensure that industrial activities are sustainable, tools that can be applied from the very beginning of the design process through to the final stage of manufacture. This paper will consider some of the tools and show how CIAM's work is helping to bring sustainability back to the North-East.

1. INTRODUCTION

The North-East of England has long been associated with manufacturing and industry. A number of major manufacturers such as Nissan, Vickers, Black and Decker and so on, are all located within a few miles of each other. Increasingly, however, the manufacturing industry has been struggling. The strength of the pound and the expense of the local workforce have forced companies to cut costs. Many companies are able to pass the problem down to their suppliers, often "Small to Medium Size Enterprises" (SMEs). The SMEs themselves are then faced with the problem of being able to supply goods at a price the customer is willing to pay, whilst retaining an ability to make a net profit rather than loss. Increasingly, therefore, many SMEs have become worried about sustaining themselves in the immediate short-term, rather than sustaining the world "for generations to come".

The World Commission on Environment and Development (WCED) was commissioned by the General Assembly of the United Nations "to consider ways and means by which the international community can deal more effectively with environmental concerns" and "to help define … a long term agenda for action during the coming decades, and aspirational goals for the world community" (1). The Commission's final report, "Our Common Future" (also known as the Brundtland Report after the chair of the commission, Dr Gro Harlem Brundtland) was published in 1987 and defined sustainability as seeking "to meet the needs and aspirations of the present without compromising the ability to meet those of the future" (2). Continuing this theme in her 1999 address to the World Business Council for Sustainable Development, Brundtland stressed that "the responsible manager has the desire to look ahead with a perspective going beyond your present individual tenures" (3).

Both the Brundtland Report and Brundtland's address make one important assumption, however – that one has a choice in how one is able to "meet the need of the present". Whilst the tools may exist to provide such a choice, many industries in the North-East, and indeed, throughout the world, are not aware of this, and therefore are only able to rely on a vision of today, rather than of tomorrow. Many managers may have the "desire to look ahead", but do not feel that they have the financial or personnel resources with which to go beyond their "present individual tenures". Instead they feel they can only afford to address the more immediate issue of keeping the company afloat. This paper will argue, however, that for many manufacturing companies in the North-East, it is precisely by addressing the problems of today that some of tomorrow's problems will be solved. It will discuss a number of design tools that the Centre for Industrial Automation and Manufacture (CIAM) can offer to industry and show how these tools can be used in the struggle to create "industrial sustainability" for today and "Brundtland sustainability" for tomorrow.

2. SUSTAINABILITY AND INDUSTRY IN THE NORTH-EAST

The traditional perception of "sustainability" is that of the Brundtland Report. It relates to the environment – to reducing pollution, emissions and so on. However, to many companies in the North-East, sustainability is more than a long term aim and relates to business as it occurs now. Should a company not be able to sell to its customer(s) at a profit, then it is the business that becomes unsustainable, often resulting in increased unemployment. Brundtland quotes Indira Ghandi as saying "Poverty is the greatest polluter" (3) and this can certainly be extrapolated to industrial opportunities – where these companies are struggling purely to survive, they may often feel that they cannot afford the time or resources to move to the higher technology – but increased expenditure – of equipment that creates less waste and is more efficient. Such companies are now being forced to act, by European legislation and British taxes.

Increasingly, these regulatory frameworks are coming into place, with the intention of penalising companies that are not making attempts to move towards Brundtland sustainability. These attempts – such as cutting emissions, using less harmful chemicals in processes, decreasing energy use and so on – may require significant financial commitment to new technology. It may therefore be thought that Brundtland sustainability and industrial sustainability are likely to lead in opposite directions and that governmental policies may perhaps not be the best tool for accomplishing either. However, imposition of the regulations has made industries consider some of the issues regarding sustainability – issues that, if addressed appropriately, could allow them to save considerable amounts of money. One specific example of this is a Teaching Company Programme currently being run between the University of Durham and Terra Nitrogen, an ammonia production facility in Billingham. The programme was initiated by Terra Nitrogen in response to the rising costs of energy use – directly through the price of fuel and indirectly through regulatory penalties – to investigate methods of reducing the energy required throughout the ammonia production process.

The BBC "Reith Lectures 2000", on the subject of "Respect for the Earth", addressed some of the sustainability issues from both sides of the "regulatory fence". Chris Patten, Commissioner for External Relations within the European Union , addressed the notion of governance for sustainable development. He linked "sustainable development" with "good governance" and "democracy" and went on to say "I think the challenge for democracies is to convince people today in the developed countries ... that success isn't just about extending appetites" (4). Thus the aim is not to manufacture more, but to manufacture better. He continued by saying that, for a government, "sustainable development is about much more than environment policy defined in terms of departments, ministers and white papers. It requires a mosaic of institutions, policies and values". In suggesting that "The pursuit of sustainable development requires changes in the domestic and international policies of every nation" (2) the WCED clearly agreed. Indeed, also in the Reith Lectures, Dr Brundtland confirmed that "sustainable development cannot work without good governance" (5) – the form in which this governance should express itself – the "institutions, policies and values" – is not made clear however.

With Brundtland sustainability in mind, the next EU Framework Programme, for research throughout the EU, is understood to want to reduce greenhouse gases and pollutant emissions, to assure the security of energy supply, to change energy consumption, and to advance renewable energy sources (6). All of these, however, are completely reliant on the existence of industry. Without industry to provide materials and components for electricity generators, how can energy supply be assured? Without manufacturers of electrical, electronic components, materials, etc. how can renewable energy sources themselves (such as turbines, PV modules etc.) be manufactured? Lord Browne of Madingley (then Sir John Browne), Chief Executive Officer of BP Amoco, presented his part of the Reith Lectures from the side of businesses. As CEO of the world's third largest oil corporation, this may seem surprising. However, he stated that "Business is not in opposition to, but has a fundamental role in delivering sustainable development" (7) asserting that "business needs sustainable societies in order to protect its own sustainability. ... Most want to do business again and again over many decades". In the drive towards sustainable development – towards the use of renewable resources, towards energy efficiency and conservation – there must also therefore be a drive towards sustaining industries. Thus industrial sustainability and Brundtland sustainability are intricately linked and failure to ensure industrial sustainability seems likely to immediately preclude the realisation of Brundtland sustainability.

3. DESIGN TOOLS FOR INDUSTRIAL AND BRUNDTLAND SUSTAINABILITY

The on-going nature of the design process throughout a product's lifecycle (initial concept generation – detailed design and embodiment - manufacture - product replacement - initial concept generation ...) ensures that any number of tools can be used successfully in individual parts of the cycle. This section will investigate a number of design tools, however, that show potential in affecting sustainability throughout the whole process. Each of these tools has been developed (to a greater or lesser extent) and applied successfully by CIAM within industries in the North-East.

Quality Function Deployment (QFD)
The origins of QFD as a means of quantifying customer requirements are to be found in Japan in the late 1960s. Japanese engineers were looking to take the thoroughness they had developed in manufacturing, into the manufactured products. QFD has since spread from Japan, into the USA in the 1980s and belatedly into Europe. However, the semi-statistical relationship encouraged by QFD, whilst enabling prioritisation, has also lead to an apparent

complexity and slowness which strongly hampers its spread in the UK.

The design method encourages a team approach to New Product Development, through the build up of a set of matrices. The matrices help design teams to define the customers' requirements, and to map out the relationships and relative importance of both how those requirements are translated into the application of technology and how they are controlled on the manufacturing shop floor. In this way it can be ensured that the customer requirements are demonstrably achieved. Where the approach can be adapted to fuse the "Voice of the Customer" (expressed through perception and feel) with the "Voice of The Engineer" (expressed through functionality and measurement), the objectives are more likely to be realised.

This process is obviously a step towards ensuring industrial sustainability through the accurate understanding and application of the needs of the customer with relation to a new product. Brundtland sustainability can therefore be attained indirectly through the sustainability of the company. However, it can equally be attained directly by recognising the needs of a wide range of "customers". For example, Governments can be customers through environmental legislation, individual customers can reflect this in their request for recyclable products, and suppliers of materials can require that environmental friendly components and processing are important within the supply chain. In his contribution to the Reith Lectures Lord Browne even recognised his own employees as customers:

> *The people who make up and shape society are the same sorts of people who work in companies like my own. ... People want to work for something they believe in ... and to make a contribution to the progress of the world in which they live. And if business is to succeed it has to offer them the opportunity to do just that.*
> *That's why our very commercial targets now themselves embrace environmental and ethical objectives. They do so, partly because our employees demand it. And our customers demand it too.*

QFD presents a process that allows the incorporation of these "customer requirement" from the very beginning of the design process and, as such, can be a very useful tool in promoting both industrial and Brundtland sustainability.

3-D embodiment
The rapid improvements in three dimensional CAD software have proved invaluable to many companies in the North-East. The ability to design, and to visualise, in three dimensions produces a number of benefits. A few of the more obvious ones are listed below.

- Embodying the design into a three dimensional visualisation allows an understanding of the product by all members of the design team. This should preclude any errors arising from uncertainties about the design prior to proto-typing.
- The visualisation can be used to promote involvement by those involved in the actual manufacturing process, allowing them to point out any problems that might arise, or to propose simpler manufacturing methods for achieving the same ends.
- Modern software often permits the direct export of CAD models to rapid prototyping equipment, removing the opportunity for discrepancies between the intended design and the prototypes.

D002/002 © IMechE 2002

• The visualisation can be used as a market research tool, to investigate the "Voice of the Customer" and ensure that the proposed product does in fact match the specification of the customer.

The combined effect of all these benefits is a reduced number of design iterations, therefore the use of less materials and time and so on. Resource expenditure, whether that be financial, human or material, is therefore kept to a minimum prior to manufacture. Such an efficient use of resources clearly fits in with both Brundtland sustainability and industrial sustainability.

Design for Assembly (DFA)
DFA methods are well established and are predominantly based on part reduction using Boothroyd's design "rules" (8), combined with ensuring ease of handling and ease of insertion of components. Many of the methods/tools available, however, are computer-based, and therefore limit the interplay between the designer at the computer, and the rest of the team which may include manufacturing engineers, marketing, finance, management and so on. Whilst this approach has proved successful in many cases, the lack of interplay prevents the DFA methods from perhaps reaching their full potential

A team-based methodology for DFA has been developed at CIAM and has been widely practiced in North-Eastern industry, due to the extensive student-company interaction that the CIAM has developed. This methodology has been presented in detail by Appleton and Garside (9), but, in order to consider its effects on sustainability, the results of some case studies must be considered. Appleton and Garside (9), quote three different projects, where part counts were reduced by an average of 77%. By reducing the number of parts, the time taken to assemble the parts is also reduced, in one study from 13.7 minutes to 4.47 minutes. In a more recent project the financial results were particularly impressive. It is expected that the new component design, arrived at through the DFA exercise, will save around £0.75 on each of the 50,000+ components sold each year. The cost of re-tooling is likely to be around £10,000, giving savings of £38,000 p.a. after the 3 month payback period.

The impact of Design For Assembly on industrial sustainability is evident. Less time spent assembling products implies lower labour (and general overhead) costs per product. The process often has further advantages such as lowering material use, or allowing cheaper materials to be used, but these are by-products of the process rather than, necessarily, specific aims. The resultant cost savings can either be used directly to increase profits, or to lower the price to ensure competitiveness. Either result will aid the long term life of the company. The Brundtland sustainability, on the other hand, can be found primarily in the increased efficiency of use of general overheads, including, for example, lower electricity and space-heating usage per manufactured product.

Design for Manufacture (DFM)
One of the results of achieving a reduction in parts through DFA is often an increase in component complexity, leading to increased manufacturing costs and potentially off-setting any savings made at the assembly stage. DFM is therefore often thought of as a natural extension of DFA and was similarly progressed by Boothroyd (8). CIAM has again adapted DFM and developed a team-based approach which has consequently been applied in numerous companies in the North East (10). The basic characteristics of this process are: eliminate unnecessary component features; enable standardisation; simplify the manufacturing

process route; improve material efficiency; and simplify tooling and equipment.

The consequences of this design methodology for sustainability – both industrial and Brundtland – are quite clear. Manufacturing is often a vast user of energy (for example, the burning of gas providing molten metal, or electricity for powering tools), thus reducing the complexity and time of the process can decrease cost to the company of that energy and, at the same time, reduce the cost to the environment of the production of that energy. The improvements to material efficiency have exactly the same results.

Lean Manufacture

The expression "lean manufacture" came into usage in the late 1980s and early 1990s to help explain the origins and development of Japan's post war development of manufacturing practice, as opposed to the earlier USA model of Mass manufacture. In this context, the term "Lean" can be considered to be an analogy with meat. Lean meat is the opposite of thin or starved meat and should have no fat. Instead, the meat should consist of muscle tissue, supple and even, strong but flexible.

Lean manufacture can be characterised by four aspects: layout and product flow; changeover characteristics; maintenance; and waste elimination. Each of these has an emphasis on the contribution to on-going improvement by the whole body of the workforce, rather than being channelled through a small number of specialists. Although they are aimed at achieving cost reduction, and hence industrial sustainability, these aspects are also applicable to the pursuit of Brundtland sustainability. For instance, layout and flow encourage the efficient use of space; changeover emphasises the effective use of equipment; and maintenance looks at achieving continued and enhanced use of existing equipment before being replaced. The last of the four, waste elimination, is particularly relevant to Brundtland sustainability. Its aim is to identify and remove the seven wastes of overproduction, waiting or delay, transportation, inventory, inefficient processes, rejects and rework, and inefficient body movements.

Quality Improvement

CIAM has carried out several projects focussed on quality improvement exercises, including "Six Sigma", "Total Quality Management" (TQM) and "Statistical Process Control" (SPC). This paper is not an appropriate forum to discuss these methods in any more detail than to say that they are different (although arguably complementary) approaches in the continuous struggle to keep quality standards high. Whereas the tools described previously aim to use less resources to create the products, quality improvement tools concentrate on lowering the conversion rate of potentially useful resources into reject – and therefore often scrap – goods.

The projects have mainly related to improving a company's response to the true requirements of its customers. The practical realisation of this is gained in the improvement of technical processes by establishing repeatable and reliable measuring systems, establishing the current level of performance, identifying opportunities for improvement, implementing improvements and establishing new control regimes to embed the improved process in the company. This is equivalent to the standard Six Sigma approach of DMAIC - Define, Measure, Analyse, Improve, Control.

The impetus behind such an approach is that if quality can be improved then the requirements of the customer (who might be internal or external) will be met more often. If this occurs then it becomes more likely the company will keep its current business and will possibly gain more business. Equally, the reduction in the manufacture of defective goods will minimise the waste of time and resources through inspection, rework, warranty and so on, and will thus help lead towards Brundtland sustainability.

4. CIAM, STUDENTS AND INDUSTRIAL SUSTAINABILITY

The design methods described above have been built on and adapted by CIAM through sustained industry-university interaction over several years. This interaction is crucial for the continued development and extension of the tools, and is evidently beneficial for both CIAM, in advancing the learning and understanding of the processes, and for industry through their improved sustainability. A further benefit for CIAM, however, is that these methods are all used for the teaching of final year manufacturing and design students, through two-week industrial placements, or one day in-company teaching days. Thus it is the students themselves that are most instrumental in bringing about sustainability within the companies. A complete discussion of this method of teaching is to be presented at the 3rd Global Congress on Engineering Education in August (11), detailing the considerable educational benefits that this brings to the students, and the financial benefits that can be brought to the companies. It should be noted that it is this kind of student/industry project that has formed the basis of the case studies mentioned in some of the design methods above.

Clearly this approach has an extensive impact on the companies. One North Eastern company has made it quite clear that they would not exist today if it had not been for CIAM using some of the design methods on their products and manufacturing. Jim Summerbell, Technical Director of Mechetronics with whom CIAM has a long standing relationship, has similar sentiments, stating that the students are able to come up with some "bloody good ideas", even to the extent that the company gets more out of the students' work than the students themselves. Whilst this may be arguable, given the education received by the students from their time in industry (see (11) for a full discussion), it is evident that the companies themselves feel that they are being helped towards both industrial and Brundtland sustainability.

Sustainable development also comes that little bit closer to being achieved through the background the students during their projects. The mind-set and analytical approach to cost and waste reduction with which the students are equipped provides effective training for the implementation of similar sustainability exercises in any business atmosphere. In effect, by having students carrying out the sustainability projects, using the tools described in this paper, the increase in and proliferation of the pursuit of sustainability is ensured.

5. SUMMARY AND CONCLUSIONS

The accepted definition of sustainable development, coming from the Brundtland report, is the ability "to meet the needs and aspirations of the present without compromising the ability to meet those of the future" (2). Traditionally the emphasis has been on the latter half of the statement, but this paper has suggested that for many companies in the North-East of England it is meeting the needs of the present that is the immediate challenge.

The Centre for Industrial Automation at the University of Durham has carried out considerable amounts of work, often through undergraduate projects, with companies in the North-East and has helped many of them to take the first steps to industrial sustainability. Tools such as QFD, 3-D embodiment, DFA, DFM, Lean Manufacture and Quality Improvement have all resulted in significant improvements in manufacturing and design processes, such as savings in materials, employment costs and overheads, lower energy usage and reduced wastage. The consequence of these improvements, although perhaps not intentional, is a movement towards sustainable development as defined in the Brundtland report. An additional benefit in the way these tools are applied is their use in the education of many young engineers and the consequent potential for their application wherever the students find themselves in their later career.

Evidently the road to sustainability is not an easy one, but it is one onto which companies have been forced, through EU and UK legislation. Many companies in the North-East have, however, discovered a number of means of travelling along this road, through the tools that CIAM have made available to them. The additional benefit to these companies is that in making use of these tools they find themselves on the road to both Brundtland sustainability and industrial sustainability.

REFERENCES

1. **Brundtland, G. H.** (1987) Chairman's Foreword, in *Our Common Future*, World Commission on Environment and Development. Oxford University Press, UK.

2. **World Commission on Environment and Development** (1987), Our Common Future: Oxford University Press, UK.

3. **Brundtland, G. H.** (1999) Our Common Future and Rio 10 years after: How far have we come and where should we be going?, Address to World Business Council for Sustainable Development, 4 November 1999.

4. **Patten, C.** (2000) Governance. in *BBC Reith Lectures 2000: Respect for the Earth*, BBC Radio 4, April 12th 2000.

5. **Brundtland, G. H.** (2000) Health and Population. in *BBC Reith Lectures 2000: Respect for the Earth*, BBC Radio 4, May 3rd 2000.

6. **McKinlay, C.** (2001) Sustainable Development and Global Change in the Next FP, (Presentation to Durham University), UK Research Office.

7. **Browne, J.** (2000) Business. in *BBC Reith Lectures 2000: Respect for the Earth*, BBC Radio 4, April 26th 2000.

8. **Boothroyd, G., Dewhurst, P., and Knight, W.** (1994), Product Design for Manufacture and Assembly. New York: Marcel Dekker.

9. **Appleton, E. and Garside, J. A.** (2000) A Team-Based Design for Assembly

D002/002 © IMechE 2002

Methodology. *Assembly Automation*, 20(2): pp. 162-169.

10. **Appleton, E. and Garside, J. A.** (2000) A Team-Based Design for Manufacture Methodology. in *ASME 2000 Design Engineering Technical Conference*. Baltimore, Maryland, USA.

11. **Short, T. D., Garside, J. A., and Appleton, E. A.** (2002) Education and the Engineering Student: A Marriage Made in Heaven? (accepted for oral presentation). in *3rd Global Congress on Engineering Education*. Glasgow Caledonian University, UK, 30 June - 5 July 2002.

Standardizing sustainability

R VALENTINE
Department of Computing and Applied Sciences, Suffolk College, Ipswich, UK

ABSTRACT

This paper aims to investigate the 'sustainable' industry by conducting an environmental analysis. From the analysis, key environmental drivers are used as a platform on which to build a future direction of 'sustainable' development for design and manufacture. The paper proposes political support for companies by certifying that such development would conserve fossil fuels, coal, oil and gas, by 'sustainable' practice. Additionally, as trends show a decline in manufacturing output, this political development considers how businesses might achieve a competitive advantage in the growing and developing 'sustainable' industry.

NOMENCLATURE

BSi	-	British Standards Institute
CCRI	-	Climate Change Research Initiative
DEFRA	-	Department for Environment, Food and Rural Affairs
DETR	-	Department for Transport
DTI	-	Department of Trade and Industry
EPA	-	Environmental Protection Agency
EPSRC	-	Engineering and the Physical Sciences Research Council
EUFORES	-	European Forum for Renewable Energy Sources
MAFF	-	Ministry of Agriculture, Fisheries and Food
STI	-	Sustainable Technologies Initiative
SMS	-	Sustainable Management System

1   SUSTAINABLE INDUSTRY

1.1   Political

1.1.1   *Sustainable technologies initiative*

Within this Initiative, £18 million of total funds is available for research and development over the next five years, for improving sustainability in UK business. Within the STI is a LINK programme. Launched in November 2000, it attracts £10 million of the total funds equally from the DTI and EPSRC to support research projects between industry and academia that develop sustainable technologies at the *design stage*, as opposed to technologies that treat end products for environmental unfriendliness. The LINK programme will supply 50% of a project's cost. A Parallel programme also exists for the development of similar technologies without an academic partnership. The DTI offers funds of £5 million for projects that are match funded by industry.

1.1.2   *Renewable energy*

The Government is showing increasing commitment to the sustainable industry by proposing targets for the UK's electricity need. By the end of 2003, 5% of the need should be met by renewable energies and by 2010, 10%. Conditionally, such targets should be achieved only through a cost to the consumer, which is acceptable. Furthermore, these targets may prove to be only the beginning of the emerging sustainable industry by; *"Whilst this is an ambitious target, it is not an end in itself. I do not want to see renewables stop at 10%. I want to see a strong, world-beating industry develop in the UK."* (Peter Hain, The Minister for Energy and Competitiveness in Europe.)

In order to meet these targets, the Government announced in March 2001 that over £260 million of funding over the next three years in support of renewable energy technologies. This inward investment is derived from five sources: £50 million National Lottery money mainly for offshore wind and energy crops; DTI's £55 million for the Government's enhanced renewable energy research and development programme; DTI's £39 million support for offshore wind, announced by the Prime Minister in October, 2000; £100 million for green energy, including solar PV, wind and wave, announced by the Prime Minister in March, 2001; and £12 million from MAFF in grants for planting energy crops. (www.dti.gov.uk/renewable)

To investigate the viability of photovoltaic (PV) solar energy technologies, the Government Industry Photovoltaics Group was formed in mid-2000. The Group was formed following a meeting the President of BP Solar, Harry Shrimp, the then Energy Minister, Helen Liddell and the Environmental Minister Mr Meacher. The Group, working confidentially with other relevant industries, shall appraise the economics and future use of PV in the UK. This does not suggest that PV will supersede other renewable technologies, but merely that the Group shall consider only how PV could be implemented effectively. In a report, the Group recognised the growth of Renewables Strategy

Two US climate centred initiatives were submitted to Congress, in February 2002, each attracts $40 million in President Bush's 2003 budget. The Climate Change Research Initiative (CCRI) shall complement the US Global Change Research Program (USGCRP), which has been operating for over a decade. The new strategy to reduce 'greenhouse gas intensity' by 18% by

2012. In the Science journal (2002) it is explained that businesses will be encouraged by \$4.6 billion in tax credits over the next 5 years for the use of renewable energy sources.

1.1.3 Environmental standards

In order to protect the environment against the adverse impact of business activity, the British Standards Institute (BSi) issued environmental standards in 1996. Companies may develop an environmental management system, by conforming to BS EN ISO 14001:1996. This International Standard helps organisations to formulate a policy, which would assure adherence to environmental legislation. Indeed, through this Standard and related standards in the BS EN ISO 14000 series, there should be political encouragement for companies to adopt them, but is this enough? At present this standard series is under review by the BSi. The review aims to utilise any similarity between EMS and the Quality Management Systems (BS EN ISO 9000:2000) by streamlining the auditing process, that organisations undergo for certification into a common audit, as opposed to two individual ones. At a time when institutions are considering the impact of standards on organisations, standardising on 'sustainability' might be considered also.

The formation of the Kyoto Protocol has established targets for reducing CO_2 emissions. This will have a positive effect on the development of sustainable technologies, as they can produce energy with no CO_2 emissions. Furthermore, the 1997 draft EU Renewables Directive would encourage the Member States to develop or implement technology that produces electricity from renewable sources.

1.2   Economic

In Section 2.1, an appreciation of the legislative growth in both renewable technology and technology which sustains our ecological system was given. However, the present cost of generating electricity without fossil fuel is expensive. The PV Government Industry Group (2001) prescribed the action to remove barriers to implementing solar energy technologies as, *"these measures alone will not be sufficient to stimulate large scale adoption of PV in the UK because, at present, the economies are not sufficiently attractive."* Furthermore, an illustrative cost of installing a PV system on the roof of an ordinary home was given as between £8500 and £15000, which would supply 50% of an average household 2kW peak. It is estimated that PV electricity would have closer economic comparability by 2015 to 2020 when fitted as part of a new domestic building.

For businesses, the economics of installing PV technology are challenged further by corporate taxation, because it is classified as a building product, short term depreciation is not permitted. Conversely, renewable fuel such as energy crops is categorised as energy plant. Economic viability could be improved yet, through recommendations of capital grants and personal tax allowances for a domestic roofs programme and Enhanced Capital Allowances for businesses. However, PV use in the UK will not benefit from the light levels of Japan, which are a third greater nor Los Angeles where they are double. Northern European countries, such as, Germany that has a 100 000 domestic roofs programme, have approximately similar levels

1.2.1 *Manufacturing sector*

The sustainable industry could inject a significant amount of trade into the EU's economy. A report by EUFORES (1999) estimates that with continued growth in the industry, over 900 000 jobs would be created in Europe by 2020. These jobs would be divided between biomass fuel production and renewable energy with 515000 and 385 000 respectively. This would represent 500 additional jobs for the UK, by 2020. However, a PV-UK Strategy document predicts 19 000 jobs by 2010.

An indication of UK's economy can be gained from the growth in GDP, shown in Figure 1. The graph shows a downward trend, which in the latter quarters, is attributed to a weakness, *"driven on the output side by a manufacturing sector in recession and perhaps slightly less robust services growth, on the expenditure side by falling trade and investment and on the income side, weak profits,"* (www.statistics.gov.uk).

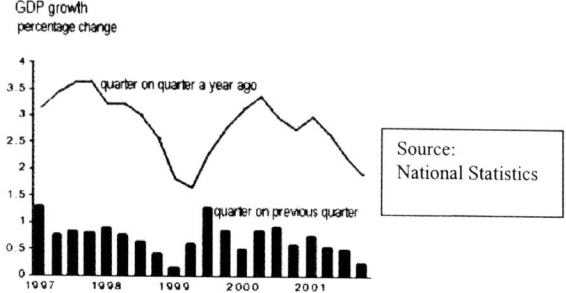

Figure 1 - GDP Growth in the UK

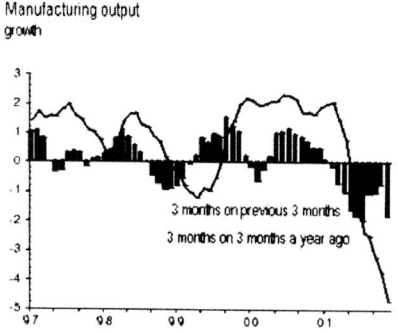

Figure 2 - Manufacturing Output in the UK

D002/023 © IMechE 2002

The decline in UK manufacturing would have been disguised in the GDP figures by the growth in the service sector. The trend in Figure 2 shows that manufacturing has been falling since the recovery in 2000, but the fall in 2001 is sharpest recorded since the recession in 1991. Indeed a global slowdown is feared as the three largest economies, Japan, United States and Germany, all experienced GDP falls in the third quarter of 2001. Additionally, stock markets all over the world have experienced large falls in the past year, particularly in technology. Valentine (2001) highlights the challenges, which British manufacturing face in terms of intensifying competition brought about by globilsation. The increase in the rate of new product development coupled with the flexibility of staff needed to help keep pace with technological change as a result of globilisation challenge company efficiency and responsiveness.

1.3 Sociological

An indicator to how people feel about the environment may be deduced from where they vote with their feet. From public attitude surveys conducted in England in 1993, 1996/7 and 2001 regarding *'Personal action taken on a regular basis'* (DEFRA, 2001), an upward trend is shown for action taken, which has a positive impact on the environment. Recycling of materials over the 8-year window is recorded as an increasing action; in 2001, over 50% of the respondents had recycled paper, 42% had recycled glass and 30% had recycled cans. As well as a reduction in the use of both garden pesticides and cars for short journeys, the public's response showed a steep trend in the use of low-energy light bulbs from approximately 18% in 1993 to approximately 33% in 2001.

It is important to consider the effect of local authority systems on the society's attitude and ability to recycle household waste. In 1983/4 the amount of total household waste recycled in England and Wales was less than 1%, in 1999/2000 this amount had risen to just over 10% or 2.7 million tonnes of recycled waste (DEFRA, 2001). The increase in recycling would reflect the growth in recycling systems that have been put in place by local authorities or organisations working with them, which have made it more convenient for the public to recycle. Through government intervention, society could be given an opportunity to act on its attitude towards the environment and, with such support, this positive environmental attitude should proliferate. To that end, the Waste Strategy for the UK, which was issued in 2000, has an objective is to recycle or compost 25 % of household waste by 2005. From this a strengthening of the recycling infrastructure from further government support could be expected together with a growth in society's participation in protecting the environment.

Indeed as society's environmental conscience is challenged increasingly, the psychological effects within consumerism, particularly household products that the UK's Waste Strategy has targeted for recycling, could become increasing more powerful. Will the consumer conscience go one step further than recycling household products and put at the top of its shopping list those products that have been designed and manufactured by sustainable practices?

The sphere of psychology in the consumer industry could be described as a complex blend of science and art, ranging from customer flow and product in-store placement to the mighty impact of graphics and colour of leading brands. However, if 'sustainable products' were to compete commercially as either a separate or new range, would the psychological side need to be marketed carefully? Alternatively, minimum disruption to consumer psychology of existing products' packaging, should they become greener or 'sustainable', could be discreetly by a symbol. This would recognise the sensitivity of packaging protection, which many leading food and drink companies have entered disputes to defend. Consequently, the industry was prompted to create an initiative to control, by arbitration, such mimicking activities as shown by *"...the client-driven code of conduct on look-a-like packaging set up by the Institute of Grocery Distribution was intended to curb the practice,"* (Mistry, 1996). All-in-all companies might need to consider whether investment in sustainability would give them a product that could appeal to a developing market and then how should the packaging reflect it.

1.4   Technological
1.4.1   Renewable solar electric power
Over the past decade the PV market has grown considerably PVGIG (2001) and with political and government funding, it is set to continue expanding. PV technology can be not only used in both off-grid and grid connected environments, but provides electricity without producing CO_2. In terms of global share in photovoltaic cell production, Japan leads the world with 40%, followed by the USA with 30.4%, the UK has 0.5%.

Developments in Building Integrated PV (BIPV) are being made in the construction industry so that BIPV can form part of the building fabric, replacing glass or brickwork. With continuing development, it is estimated that BIPV would become standard building material in 5 to 10 years.

1.4.2   Renewable transport fuels
The distribution of manufactured goods is a primary activity in the value chain and would, therefore, be affected by sustainability in the fuels that are burnt when delivering goods to the customer. From a DETR (2000) research paper, it was disclosed that the technology to produce biodiesel and bioethanol is now available. The technology achieves this from conventional feedstock, which from the wider perspective of energy balance, was said to be a preferred option. Additionally, research is being conducted in the production of bioethanol from lignocellulosics and should be graded as a medium term option. Importantly, fuels such as bioethanol can be used in conventional engines now as a blend. Bioethanol has attractiveness in that it can be used also as a fuel for fuel cell technology. The production of biomethanol or use of other vegetable oils as fuels offers little commercial interest. Fuel cell technology was placed as a longer-term renewable option. Its availability would not be for between 10 and 20 years.

1.4.3   Energy Efficient Products
The Environmental Protection Agency (EPA) in conjunction with engineering companies such as Panasonic, RadioShack Incorporation and Vtech communications, have released cordless telephones, answering machines and cordless combination unit, which because of their very low power consumption, bear the Energy Star label. These products were released at a consumer electronics show in Las Vegas in January 2002 and augment an existing range, which includes televisions and VCRs. A performance specification has been established by the EPA, which stipulates an energy operating level, for the telephones this is approximately a third of existing models. As an example of the impact this could have on the environment, *'If all cordless phones, answering machines and combination units sold in the US in the next ten years have earned the Energy Star, Americans would save approximately, $4 billion on their electricity bills'* and also

prevent greenhouse gas emissions associated with the generation of that energy *'equivalent to the emissions of more than one million cars.'* (http:yosemite1.epa.gov)

2   POLITICAL DEVELOPMENT

2.1   Future Effect of 'Sustainable' Practice on UK manufacturing

From the sustainable industry analysis, in Section 2, it is shown that significant resource is being pumped into researching and developing technology that will generate energy without harming the environment by depleting finite fossil fuels or releasing greenhouse gases, such as CO_2. Although some of the technology is available now, commercial availability is likely to be in 3 to 5 years. However, it is important to consider the effects of these renewable strategies and draft Directive on manufacturing, which is experiencing difficult times. To do this, it is necessary to view 'sustainability' in its widest terms or, perhaps, as an industry that is emerging and set to grow considerably over the next two decades or more. Arguably, the 'sustainable' industry is more than just developing renewable technology or being environmentally friendly. For design and manufacture, it promotes an ethos of 'sustainable' practice throughout its value chain through to their customers with a product or service. Henriques (2001) cites a definition of 'sustainable' development from the Brundtland report as, '…meets the needs of the present, without compromising the ability of future generations to meet theirs,' but concedes the difficulty in what it might mean in practice. This paper seeks to apply 'sustainability' to all activities in the value chain, in particular, design and manufacture, thereby giving businesses the opportunity to market 'sustainable' products. The analysis in Section 2 shows society's increasing commitment to environmental protection and consumer growth in environmental products. It is believed that with suitable marketing from pertinent stakeholders, an opportunity exists to encourage and support products made by 'sustainable' practice. However, such support needs to be political, to assure the discerning customer that the product or service supplied meets a recognised standard. Indeed, any such policy could give design and manufacturing companies that practise 'sustainability', a competitive advantage in this developing industry. Conceivably, a threat to UK manufacturing exists, on the basis that they compete in this global 'sustainable' industry and manufacturers in other countries do.

2.2   Value Chain

In the design and manufacturing industry any such policy should apply at all levels of the value chain including the end product itself. As the sustainable industry continues to grow, design and manufacturing businesses might have an opportunity of becoming competitive in it. Any new policy should, therefore, support those activities of the value chain, in design and manufacturing businesses, which would underpin a competitive advantage (adapted from Johnson & Scholes, 1993). Similarly, the new sustainable policy would support any business that equates to a primary activity only in its customer's value chain, because it is a supplier providing it with a sustainable resource. Ultimately, businesses would need to introduce 'sustainable' practice to their value chain by, for example, integrating renewable technologies within their manufacturing operations or developing 'sustainable' practice within its technology development support activity, such as, design and R and D.

2.3   Standards in Management Systems

In the 'sustainability' industry analysis is identified two management systems, Environmental Management Systems and Quality Management System. The rate at which companies become certified for the former Management System, BS EN ISO 14001 will depend on several factors. Companies would have to dedicate financial and human resource to gaining the Standard and consequently, seek to determine a return on their investment. When compared to the BS EN ISO 9000:2000 series, it does not currently possess the same business weight of demand where some companies will stipulate the Standard's conformity prior to trade commencement. Therefore, could design and manufacturing companies justify, in financial terms, expenditure on the 14000 series over and above that made for essential legal environmental conformance, in markets that are already fiercely competitive? For those companies that gain certification of the respective ISO 9000 series section, would also gain a competitive advantage by maximising their tendering power for potential work orders. Therefore, although this Standard is not a legal requirement, it would serve as an advantage to have it.

'Sustainable' development in the design and manufacturing activity of the value chain could be assisted therefore, by developing a similar management system. The three Management Systems could overlap in their operation as shown in figure 3, where, for example, a design and manufacturing company conforms to the practice of each. Putting in place a 'Sustainable' Management System (SMS) to a British Standard could give UK manufacturing an advantage against global competition in the growing 'sustainable' industry. Any British Standard would have to be set by BSi's Standards Policy and Strategy Committee and developed through an assigned sector committee. Comprising key stakeholders such as, industry, government and consumers, the sector committee could then set out to define a standard.

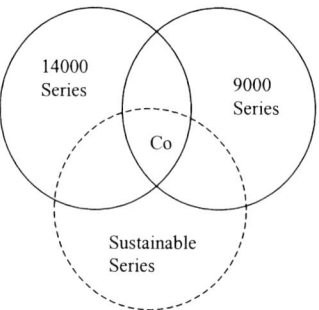

Figure 3 – Management Systems

A standard that audits a company for a SMS would need to consider in its specification, the inputs to each activity, in particular, manufacturing, but like BS EN ISO 9000, including design, development, inspection etc. In this way, companies could use 'sustainable' inputs as a basis for developing the operation within each activity.

2.4   Product Marking

A powerful marketing tool for any company's product would be a symbol of recognised quality, the BS Kite-mark is such an accolade. The Standard applies directly to a product, easily communicating to the discernible consumer that it conforms to all the appropriate standards, thereby becoming a product of quality. When the Kite-mark symbol is added to a product, with it comes the opportunity to differentiate it from other brands.   Marking or labelling of an environmental product is evident in the electronics and communications industry by the Energy Star label on those products that meet the EPA's performance specification.

In a similar way products which are designed and manufactured by sustainable procedures or procedures that utilise fuels generated by sustainable technologies could be certified to a sustainable standard and receive a symbol of recognition. By developing a British Standard of 'sustainable' performance, products could be proven or tested for adherence to it. The sector committee, in developing the specification for a performance standard, would need to establish criteria of 'sustainability'. Indeed, this represents an area of further research, as 'sustainability' is likely to be a function of several parameters. Qualitatively, a useful reflection would be of the consumption of the earth's natural resources when manufacturing a product directly. And here lies the key difference in such a Standard. Currently, existing environmental standards consider how companies dispose of waste products in manufacture by focusing on the effect its output has on the environment, whereas 'sustainability' is arguably, the focus on the effects of the inputs to manufacture on the environment! In the case of the Energy Star label, this is concerned with the consumption of any energy in order to make it function and is considered, therefore, to be under a different classification or standard, although it has very important, conserving effect on the environment. Whereas the performance standard of 'sustainability' assesses the quality of 'sustainability' of the inputs directly needed to manufacture a product. Consumers could make then a judgement on the effect of a product's energy input in manufacture on the 'sustainability' of the earth's resources.

3   CUSTOMER AWARENESS

A key factor in the success of these standards of 'sustainability' is customer awareness. Therefore, marketing their value to the customers would be an important underpinning activity to any competitive advantage.

3.1   Business segment

Any marketing campaign would need to address the two different customer bases, businesses for the SMS and consumers for the product performance standard.   The sheer scale of the 'sustainable' industry would necessitate the involvement of key stakeholders such as government, business support agencies and standardising bodies to deliver an agreed message throughout the development and growth of this industry to organisations at regional, national and international levels.   Furthermore, the type of marketing and support is likely to vary considerably depending on whether it is being addressed to large organisations or SMEs.

3.2 Public segment

Educating the general public presents different issues. Noticing a symbol of 'sustainable' quality on a product would be a relatively easy task, but conveying its full environmental implication would be more difficult. 'Environmentally friendly' products and organic products already exist so how differently would customers perceive 'sustainable'. Importantly, companies with 'sustainable' products must feel confident that their customers are fully aware of its meaning. It is this knowledge that would differentiate their products from similar products with which they compete. Consider the light bulb industry. Although low energy types have been on sale for a number of years, the traditional less efficient ones are still sold in vast quantities. Therefore, could this be the same in the 'sustainable' industry?

Consider a future scenario. A design and manufacturing company implement PV technology to drive a new product line and uses 'sustainable' practice to produce a 'sustainable' performance product, which complements an existing range of products that are manufactured by traditional methods, but operates in a different market segment. If this situation of 'sustainable' and traditional products coexist in competition, then it is likely to need significant support from the government in financing and marketing, to achieve success through development as quickly as possible.

Noticeably, utilities are helping to raise public awareness of renewable energy through their public information leaflets. In a recent publication, it was announced that *'Anglain Water has joined forces with TXU Europe to build a revolutionary renewable energy plant in Corby. The first of its kind in the UK'* (Anglian Water, 2002). Companies might wish to market 'sustainable' products in conjunction with the utilities or even greenpeace, that already have some public confidence.

4   CONCLUSION

This paper has conducted an environmental analysis on the 'sustainable' industry. Recognition is given to the considerable global investment in renewable technologies, both politically and financially. Strong indication, from this analysis, points to a future market of 'sustainability', which shall encompass not only renewable energy, but business activity within the value chain and the marketed product that is manufactured by them. However, UK manufacturing was shown to be experiencing trading difficulties through a rapidly declining level of output. Concern was paid to UK industry regarding the effect 'sustainability' would have upon it, in particular, the cost of technological integration and adherence to imposed regulation.

In order to support UK manufacturing, the development of two British Standards is proposed. Firstly, a 'Sustainable' Management System, which operates in a similar manner to the existing Systems of BS EN ISO 14001 and BS EN ISO 9000. Secondly, a product performance Standard is proposed for consideration. This standard would contain a specification that measures the quality of 'sustainability' each time a product is manufactured. Adherence to the standard would permit manufacturers to label the product, in a similar manner to the BS Kite-mark, to show the discernible consumer that it is a product of 'sustainable' quality. Both of these standards are offered as suggestions for review and further research. Their aim is to recognise politically those companies that help to protect the environment through 'sustainable' practice by developing a standard that they could use in marketing thereby gaining a competitive advantage.

5   BIBLIOGRAPHY

Anglian Water	2002	Anglian Water pioneers green technology, Watersource, D188/2/02
DEFRA	2001	The environment in your pocket, National Statistics, published by DEFRA.
DETR	2000	The role for Renewable Transport Fuels in the UK, Renewable Fuels Transport Seminar, 13-12-2000, www.dti.gov.uk/renewable/
EUFORES	1999	The Impact of Renewables on Employment and Economic Growth,
Henriques, A.	2001	Sustainability – A manager's guide, British Standards Institution, ISBN 0 580 33312 4
Johnson & Scholes	1993	Exploring Corporate Strategy, Third Edition, Prentice Hall
Mistry, B.	1996	Design Week, *"Tearing into the problem of lookalike packaging"*,23 August, 1996
PVGIG	2001	Final Report, 26 March 2001, www.dti.gov.uk
Science	2002	More Science and a Carrot, Not a Stick, vol. 295, 22 February 2002, p1439, www.sciencemag.org
Valentine, R	2001	FEA:linking theory with Practice, Engineering and Product Design Education 2001 Conference, What can Design Education do for Industry? Institution of Engineering Designers

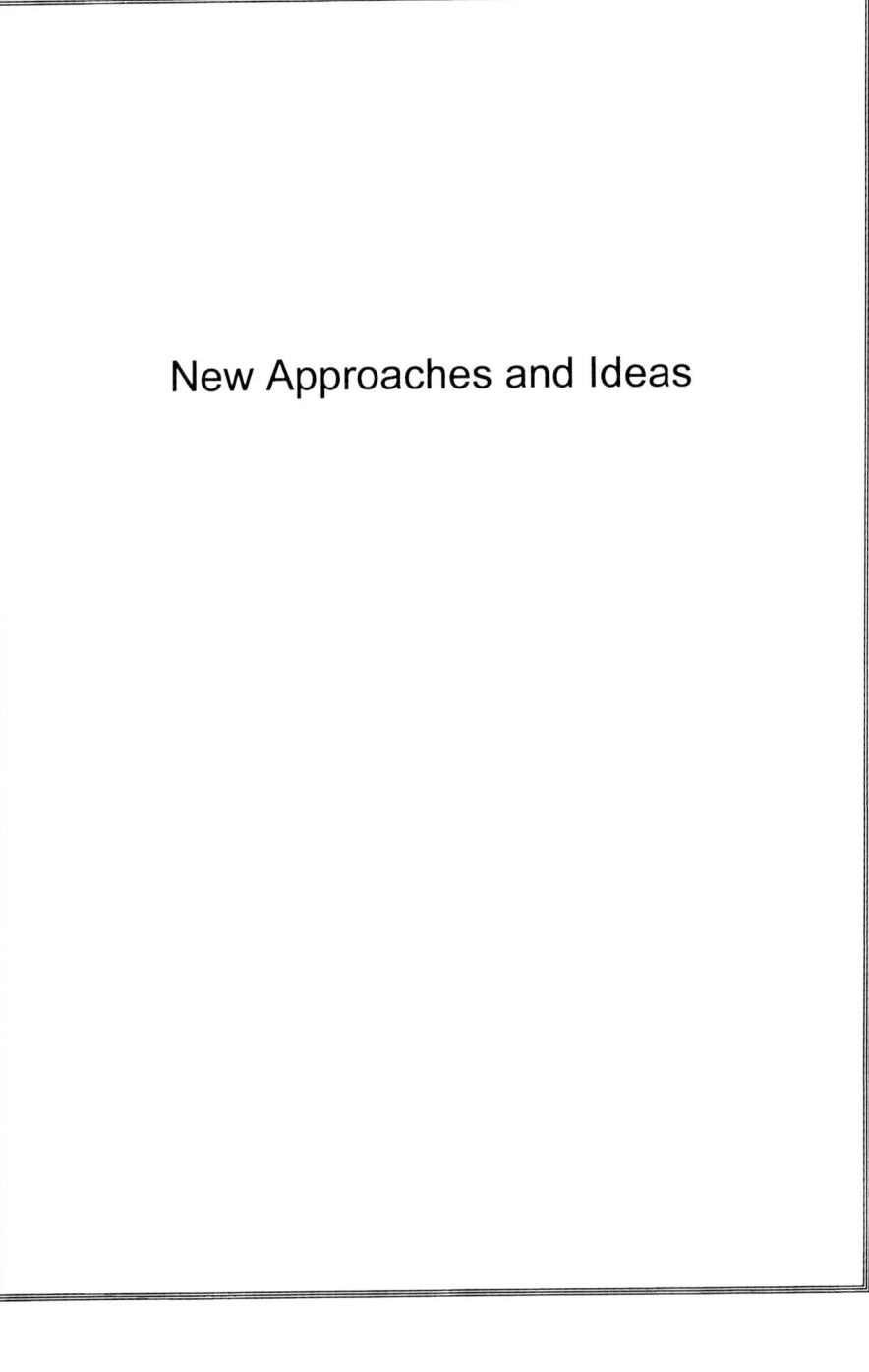

New Approaches and Ideas

Taking public perceptions of risk into account in engineering design for sustainable development – a multi-attribute decision making framework

J HARVEY, S JOYCE, and P NORMAN
Engineering Design Centre, University of Newcastle-upon-Tyne, UK

ABSTRACT

Assessment of societal risk perception is a different process compared to traditional risk assessment based on probabilistic risk evaluation methods. Where uncertainty and subjectivity exist within a decision, a structured decision making method that can capture public risk perception would help designers to progress towards sustainable product design. In this paper factors which contribute to societal risk perception are first discussed. Secondly, a multiple attribute decision making methodology is applied which could be used to objectively introduce public risk perception into design decision making. This is illustrated with an example.

1 INTRODUCTION

For design to be sustainable it must address all the issues of sustainable development: economic, environmental and societal. Of these it is the social issues which are proving to be the most difficult to incorporate into engineering design. Public perceptions of risks have been researched for some time, but a major problem has been the inability of most people to articulate their perceived risk in terms that designers, decision-makers and experts can understand and use (1).

Increasingly, societal and public concerns are forcing organizations to take a wider view of engineering design and decision-making. Some high profile cases have shown that such concerns can substantially change organizational, national, and even international strategies and force them to engage or disengage with major projects in a way that they had not anticipated or prepared for, e.g.. Balfour Beatty and the Ilisu dam project in Turkey, Shell and the disposal of the Brent Spar and so on. Such public concerns can arise for environmental and social reasons, as in these examples, or for health reasons (e.g. the BSE crisis) and can also arise when the potential outcomes and risks are not at all clear (e.g. GM foods). It is clear that public perceptions of trust and decisions about scientific issues are changing and are becoming more empowered (2,3).

High profile controversies and widely publicised scientific uncertainties and disagreements over issues like vaccination and the safety of waste incineration have had a profound impact on public confidence (4). People doubt the ability of politicians and experts to make decisions on complex issues involving scientific uncertainty. Also people have become 'sensitised' by these issues and therefore more likely to object.

2 EVALUATING SOCIETAL RISK PERCEPTION

Why take society's perception of risk into account during engineering design decision making? Traditionally in the UK, decisions have been made on the basis of expert opinion and evaluation from design engineers, project management and external regulatory and advisory authorities, such as the Environment Agency or the Health and Safety Executive. There is a normative argument that recognises the ethical reasons for the public to be represented. Risks taken by industry, for industry's benefit, may impact on society's health and safety, civil liberty and social welfare in the short term and for future generations. There is also an epistemological argument that the public's perception of the risks may represent additional information which can contribute to the decision model. In an area of scientific uncertainty experts will not have all the information and often different types of knowledge and approaches can be valuable input. This is especially true of local communities bringing local knowledge (5,6,7).

Allowing the public to participate in decision making reassures them that the right decisions are being made. From an industry point of view it allows for greater predictability as public reaction to new technology can be assessed much earlier in the process avoiding any backlash to decisions at a later, and costlier, stage (8). This offers the possibility of avoiding business decisions that are unacceptable to society e.g. the deep sea disposal of the Brent Spar (9).

3 THE ATTRIBUTION OF PERCEIVED RISK

What factors cause the public to attribute risk to an event? People are less concerned with risk probability and more concerned with risk possibility (10). Certain factors may augment the degree of risk attributed to an event by society, for example controllability, social equity, familiarity and issue salience (11,12,13). This list is not exhaustive. Many different factors are hypothesised to increase risk perception. Is it possible to capture some of these factors in order to develop indices of risk perception that can be included in design decision making? Would such a decision making process assist in the achievement of sustainable design?

4   MEASUREMENT OF RISK PERCEPTION

Risk is itself a complex issue: real events, outcomes and probabilities can be determined in many areas. Usually, the path of research has been to determine some sort of utility, U_i, where $U_i = P_i \times O_i$, i.e. some sort of product of probability, P_i and outcome, O_i for each event, and to attribute some level of rationality to the perceiver. Several theories have attempted to account for responses to personal risk (e.g. risk homeostasis theory) but explaining the sometimes extreme and not always rational responses in societal terms seems to be much more difficult. Thus, informing decision making at the design stage of any large sustainable development presents a problem.

In order to address decision making where outcomes are complex, uncertain and have potential impact upon the public, much research has gone into modelling the processes involved. In multiple attribute decision making (MADM), the decision problem is modelled in terms of the attributes of feasible alternatives and an optimal solution is achieved through a process of trade off (14). The difficulty in integrating environmental and societal issues with traditional design criteria is that they can be complex, subjective and uncertain. A fuzzy set approach allows for this uncertainty whilst also enabling an option to be evaluated (15).

In many areas of life ordinary people are confronted with linguistic terms relating to scientific measures which they are obliged to interpret; an example would include weather forecasts ("the pollen count will be high today, but medium tomorrow"). MADM may be used so those variables in the fuzzy sets can be assessed by linguistic terms instead of numerical values. The use of natural language allows a more direct representation of individuals' information in situations that are too complex or uncertain for a precise quantitative (numeric) term. The use of linguistic terms requires some ways of identifying eventually the meaning of each of the terms, and, where there is more than one person making the attribution, some form of weighting such as linguistic ordered weighted average (LOWA) (16,17).

5 THE DEVELOPMENT OF A MULTIPLE ATTRIBUTE DECISION MODEL FOR ENGINEERING DESIGNERS

It has already been established that public perceptions can be imprecise and distorted and that they can also involve issues about which the real probabilities of outcomes are not clear or sometimes not even known. Thus, measuring in this area must take into account the nature of perceived information. Fuzzy sets offer a real opportunity. This paper presents a proposed framework adapted from MADM principles to inform engineering designers about possible public perception issues.

The notion of LOWA allows us to use phrases with which people are familiar to be equated to pre-determined levels of particular issues, and to combine these to achieve a form of 'average' using a recursive process of pairwise convex combinations (18).

For an engineering design to incorporate sustainable development principles, it is crucial that public perceptions of risk and consequences are considered. In order to do this using MADM, there are several important steps:

i.   To establish the criteria [social perception indicators] which are to be rated. As well as the more obvious indicators, these must include options which describe the worst and best scenarios possible. The reason for this is to account for the problems encountered in many recent major projects, where relatively unforeseen public concerns have resulted in eventual abandonment of the projects.

The MADM methodology proposed here allows for subsequent additions to the set of criteria should the need arise as events modify. To obtain the criteria, the most obvious initial procedure is to brainstorm using experts. Experts in this context must include all the engineers and scientists who can offer any angle on the proposed project and its alternative solutions. However, it may also be necessary to consult experts in public perception of risk in order to elicit the best and worst case scenarios.

An entirely fictitious example created by the authors is used to show how such a list may be compiled; it is not meant to be complete or even realistic but to serve as a case study:

Example: *Compulsory electric control of car speeds on a highway, using embedded cables on the ground and pick-up mechanisms in cars [one option from many to be considered by designers and planners as a possible solution to traffic speeding]*

A brainstormed list of potential risks and outcomes [as perceived by the public] might initially include:

> the consequences of a power surge [Psurge]
> the possibility of electrocution [Electro]
> breakdown of system; cars out of control [Breakdn]
> level of public consultation [Consult]
> probability of [injurious] car crashes [Crashes]
> traffic jams [Jams]
> power cuts [Pcuts]
> no possibility of [emergency] stopping [No stop]
> pollution [Pollutn]
> drunk drivers [Drunk d]

The 'worst scenario' for this case study could be electrocution.

ii. Once the criteria are established, these must be normalised. This involves experts defining the anchor points which will be used in the linguistic term set. In this paper, we propose that the term set be seven units, anchored at each point and with a mid point. Although it is possible to use longer term sets [e.g. 9 or 11 points], it is proposed here that a 7-point terms set can be handled psychologically more easily than longer sets and a mid point is preferable to no mid point (19,20,).

If 7 linguistic terms are to be used, they would follow the following pattern of anchor points:

> very high   (VH)
> high   (H)
> medium high   (MH)
> medium   (M)
> medium low   (ML)
> low   (L)
> very low   (VL)

Of course, the scale could equally be 'very good' through to 'very bad', depending on the appropriateness of the high-low, good-bad etc. semantics. A 9-point scale, not

D002/006 © IMechE 2002

proposed here, would include additionally extremely high and extremely low. Less than 7 points are likely to result in the set being too fuzzy for use in the context of engineering design.

The real [not perceived] distribution of the criterion might be normal, trapezoidal, triangular or indeed one of many others (21). For the purposes of the authors' example, we shall assume a normal distribution.

With these anchor points, it is now the job of the experts to define what the linguistic operators mean in real terms. For example, we might take the criterion 'Pollutn' and the experts must then define the range of pollution which each linguistic terms means. This would start by defining the two extremes and the mid-point, referenced to the road as it currently is; to do this would mean some access to measures of pollution in the vicinity in order to express the range fully. After that, each term point can be specified as a range. In this case, although pollution is itself a complex of differing chemicals and pollutants, the experts must still either treat it as a whole, or split only into [possibly two] separate criteria to which the public could potentially relate. So the pollution levels might be defined as parts per million in the air with an equal range in each linguistic term.

The job of anchoring and defining the range for each criterion is neither simple nor speedy, but is a necessary step. The use of ranges reflects the real world and also the fact that these criteria are inherently fuzzy.

iii.   The third stage is to assign weights for each criterion. Weights describe the importance or salience of each criterion. The notion of what importance might actually mean (perceived risk, level of dread, its effects on the individual, its effects on society, etc.) has not always been well described in the literature, although the need to ascribe some rating of importance in at least a relative sense is clearly recognised. These weights reflect the membership of the set by giving a fuzzy value on the interval 0-1 for each criterion. Again, this is the job of the experts to compile, although the relative priorities might differ very considerably from those that the public might generate.

Whatever method of allocating importance is used, for example points allocation, the data are normalised so that all the criterion values in the fuzzy set aggregate to 1. Thus, the effect is to provide a type of ipsative 'score' such that a high rating of importance for one criterion must mean that others are relatively lower as a consequence; this is done to enable comparisons between sets and to reflect the reality of life where all manner of options, choices and decision must eventually be considered in such an ordinal way.

Using our electric speed control example, three sets of weights were compiled by the authors for the 10 criteria identified above. These are given in table 1

Table 1: experts weightings and aggregates for ten risk elements

	Expert 1	Expert 2	Expert 3	Agg experts	Ave experts
Psurge	.05	.30	.01	.09	.120
Electro	.01	.30	.01	.32	.107
Breakdn	.10	.02	.03	.15	.050
Consult	.20	.03	.20	.43	.143
Crashes	.10	.01	.05	.16	.053
Jams	.20	.04	.30	.54	.180
Pcuts	.05	.05	.05	.15	.050
No stop	.02	.05	.05	.12	.040
Pollutn	.02	.10	.10	.22	.073
Drunk d	.25	.10	.20	.55	.184
total	1.00	1.00	1.00	3.00	1.00

If expert 1 were considered to be more knowledgeable about for example, pollution, then his weight could be doubled relative to the judgements of expert 2 and expert 3. Similarly, if expert 2 and expert 3 were more expert about Psurges and Crashes respectively, their judgements for inclusion in the Agg experts column could be similarly doubled or even trebled accordingly. Such increases would increase the aggregated scores for several criteria, and of course the overall total for Agg experts; this would then be normalised to ensure that the Ave experts column totalled 1.0.

These weights, once assembled, remain for all the linguistic ratings, both those of experts and members of the public. The implications of these weightings is that the most important issues, in the judgement of the experts, would be how this system handles the problems of drunken drivers and traffic jams, whilst the experts consider the least important issue is whether or not drivers are prevented from stopping in an emergency- these may indeed reflect concerns currently about motorway driving!

iv. The next stage is to aggregate the linguistic ratings of, in this case, the experts. This is done by the recursive process of pairwise convex combinations (22). In the table, only three ratings for each criterion are combined, but this could be achieved with any number of ratings. As an example, for Psurge, the ratings are placed in decreasing order, i.e. Medium, Medium Low, Low. Starting with the least, Low and Medium Low are aggregated to Low, which is then aggregated with Medium to yield Medium Low.

Table 2: the expert linguistic ratings and their aggregation

	Expert 1	Expert 2	Expert 3	Agg experts
Psurge	L	M	ML	ML
Electro	VL	VL	VL	VL
Breakdn	ML	M	L	ML
Consult	ML	H	M	M
Crashes	L	VL	MH	ML
Jams	MH	VL	H	M
Pcuts	L	M	L	ML
No stop	L	L	VH	MH
Pollutn	L	L	VL	L
Drunk d	M	H	M	MH

Table 3: the weights and linguistic rating summary table

	Ave weight	Agg Linguistic
Psurge	.120	ML
Electro	.107	VL
Breakdn	.050	ML
Consult	.143	M
Crashes	.053	ML
Jams	.180	M
Pcuts	.050	ML
No stop	.040	MH
Pollutn	.073	L
Drunk d	.184	MH

If we attribute a numerical scale to our linguistic scale e.g. very high=6, through to very low=0. We can obtain an overall 'score', which may be ranked, as in Table 4.

Table 4: weightings, linguistic weights and final rankings

	Ave weight	Agg Linguistic	W x L	Final ranking
Psurge	.120	ML	.240	4
Electro	.107	VL	.000	10
Breakdn	.050	ML	.100	7=
Consult	.143	M	.429	3
Crashes	.053	ML	.106	6
Jams	.180	M	.540	2
Pcuts	.050	ML	.100	7=
No stop	.040	MH	.160	5
Pollutn	.073	L	.073	9
Drunk d	.184	MH	.736	1

At this stage, the designers may want to consider those criteria which have the highest linguistic ratings, since those are likely to generate considerable public concerns in relation to this proposed solution to the traffic problem. However, the importance of the criterion may also reflect public opinion. The final ranking in our small case study makes it very clear indeed that there are in fact three primary public concerns, much greater than the other criteria, being: the problems of drunken drivers, traffic jams and public consultation.

This method may be repeated for each option that may be considered feasible by the designers. It may also be used as a linguistic tool to allow members of the public to evaluate various options and risks directly.

6 DISCUSSION AND CONCLUSION

This method presents a way of obtaining some relative evaluation of risks by 'pooling' the judgements of experts in two ways: to establish the weightings and to evaluate risks and outcomes linguistically. Its use can them be extended to allow the public to evaluate options linguistically using the criteria and weights derived by the experts. The recursive process of pairwise convex combinations can be employed successfully to produce an aggregated linguistic assessment for each of the issues, and indeed the final linguistic column could itself be aggregated to yield an overall single linguistic if so desired (17).

When the experts are assembling the values initially which will correspond to the linguistic weights, it is clear that many measures will not contain equal intervals (for example noise as an outcome would follow a basic log scale, but may then be "biased" in discriminatory terms towards higher levels of noise). So it is clear that the scaling could be normal, lognormal, trapezoidal, triangular, etc.; this is why the scaling must be decided by experts and not by the public.

The purpose of this method is primarily to inform engineering designers of the possibilities of public perceptions and beliefs in relation to the risks involved in the various design options that they may wish to consider, in order that issues may be anticipated rather than confronted when it is too late. A survey of two major UK engineering companies (23) revealed that designers are fully aware of their responsibilities relating to the sustainability of their products but they lack appropriate tools and methodologies to allow them to bring the issues fully into the mainstream of the engineering design process. Robust decision making methods applicable throughout the design lifecycle stage are needed to ensure sustainability objectives developed early in the design process are adhered to when in conflict with traditional criteria such as cost, weight and reliability. The work reported here is aimed at alleviating this situation by formalising what is currently a qualitative, uncertain and subjective process such that a practical methodology can be evolved that is transparent and trusted by all stakeholders.

REFERENCES

1. **Slovic P, Fischoff B and Lichenstein S** (2000) Cognitive Processes and Societal Risk Taking. In: P Slovic (ed) *The Perception of Risk.* London: Earthscan
2. **Barr C** (1996) Fear Not: The Art of Risk Communication Journal of Management in Engineering 12 Jan-Feb pp18-22

3. **Harvey J and Erdos G with J Holme, T Raven, GDL Staunton, A Walton** (2002 in press) An empirical study of protein consumption and attitudes to genetically modified food. Risk Decision and Policy. Vol 7.

4. **Green Alliance** (2001) Decision making under scientific uncertainty: the case of mobile phones. June 2001 Green Alliance

5. **Pidgeon, N** (1998) Risk assessment, risk values, and the social science programme: why do we need risk perception research? *Reliability Engineering and System Safety* Vol 59: pp5-15

6. **Barr C** as ref.2

7. **Green Alliance** as ref. 4

8. **House of Lords** (2000) Science and Society. House of Lords Committee on Science and Technology Report HMSO

9. **Rice T and Owen P** (1999) *Decommissioning the Brent Spar.* E and FN Spon

10. **Carter P and Jackson N** (1992) The Perception of Risk. In : J Ansell and F Wharton (eds) Risk: analysis and management: Chichester: John Wiley

11. **Slovic P, Lichtenstein S and Fischoff B** (1984) Modelling the Societal Impact of Fatal Accidents. *Management Science* 30 (4) pp 464 -474

12. **Barr C** as ref. 2

13. **Pidgeon N** as ref. 5

14. **Sen P and Yang JB** (1998) *Multiple Criteria Decision Support in Engineering Design.* Netherlands: Springer-Verlag

15. **Zahed L** (1973) Outline of a new approach to the analysis of complex systems and decision processes. *IEEE Transactions, Systems, Man and Cybernetics* Vol 3, part 1: pp28-44

16. **Herrera F, Herrera - Viedma E and Verdegay JL** (1996) Direct approach processes in group decision making using Linguistic OWA operators, *Fuzzy Sets and Systems* 79: pp175 - 190

17. **Stoyell, L** (2000) *Fuzzy multi-criterion decision making exercise.* Unpublished Engineering Design Centre report. University of Newcastle upon Tyne

18. **Herrera F, Herrera - Viedma E and Verdegay JL** as ref. 16

19. **Oppenheim AN** (1994) *Questionnaire design, interviews and attitude measurement.* London: Pitman

20. **Valls, A and Torra V** (2000) Using classification as an aggregation tool in MCDM. *Fuzzy Sets and Systems* 115 pp159 - 168

21. **Herrera F and Verdegay JL** (1997) Fuzzy sets and operations research: perspectives. *Fuzzy Sets and Systems* 90 pp207 – 218

22. **Herrera F, Herrera - Viedma E and Verdegay JL** as ref. 16

23. **Stoyell J** L, Norman P, Howarth CR and Vaughan R, (1999) results of a Questionnaire Investigation on the Management of Environmental Issues during Conceptual Design: A Case Study of two large Made-to-Order Companies. *Journal of Cleaner Production*, Vol 7 No. 6 pp457 -464

Incorporating life-cycle cost into early product development

J-H PARK and **K-K SEO**
CAD/CAM Research Center, Korea Institute of Science and Technology, Cheong-ryang, Seoul, Korea

ABSTRACT

Life cycle concerns have been considered as a major issue of increasing importance. Recently, companies have realized that both environmental and economic aspects become a more competitive issue. This environmental and economic performance of a product includes all life cycle phases, from raw material extraction to end-of life treatment. Life Cycle Cost (LCC) has been considered to enable comprehensive cost analysis to improve economic performance. But this method is usually time and cost consuming. Therefore, there is a need for easy-to-use and approximate methods to support cost-effective decision-making in early product development. In order to incorporate life cycle cost into early product development, an approximate method for the estimation of maintenance cost as one component of life cycle cost is proposed. This method allows the designer to make comparative estimation of maintenance cost for design alternatives by considering high-level product attributes and the maintenance cost during life cycle of products. To estimate maintenance cost, the identified product attributes are used as inputs and the calculated maintenance cost is used as outputs in learning algorithm based on artificial neural networks. The proposed approach does not replace the detailed cost estimation but it would give some cost-effective guidelines of products during the life cycle and create possibly new maintenance concepts.

1 INTRODUCTION

Economic globalisation opens the door for participation and development for all into the global economic game bringing about growth and developments. The economic forces associated with these developments put companies into an always more short-term orientation. Keen competition worldwide forces enterprises to shorten their product development lead-time. New products with more and more functions and better performance at lower prices are entering the markets faster than ever. Products must be developed very quickly with little time thorough both environmental and economic analyses according to increasing life cycle concerns of products.

In engineering design research activities have been concentrated on the environmental and economic performance a product during the life cycle in order to compete in the global market. Some methods and tools have been developed to evaluate economic and ecological consequences of activities during the life cycle of a product to give decision support to designers and managers.

Determining product design is a critical task in any product development activity. By selecting a suitable design for a product, large cost and time savings can be realized. While many different objectives must be weighed when determining product design, this paper focuses specifically on how incorporating life cycle cost (LCC), maintenance cost (MC) in particular, into early product development. In the early phases of the life cycle, when a new product is considered, cost estimate analyses are used to support the decision for product design. Later on when alternative designs are considered, the best alternative is selected based on its estimated life cycle cost and its benefits.

The increasing intensity of LCC-based competition motivated a significant research effort aimed at developing methods that support product maintenance cost estimating analysis in early product design. This paper describes research to develop the estimation method for the product MC in early product design. This method allows the approximate and rapid estimation of the product MC based on high-level information typically known in the conceptual phase. An artificial neural network (ANN) is trained on product attributes and the MC calculated from historic maintenance data. The product designers query the trained artificial model with new high-level product attribute data to quickly obtain an approximate MC for a new product concept. The proposed method is to minimize the life cycle costs of products and support cost-effective decision-making as the efficient and easy-to-use method in early product development.

This paper is organized as follows. In section 2 a brief description of background of the paper is provided and related works are reviewed. In section 3, we define and identify product attributes used to estimate product MC using the ANN method in conceptual design. Section 4 describes the procedure of the calculating MC. The test of the ANN models is presented and then discussed an evaluating the performance of experiments in section 5. Finally, some concluding remarks and future works are provided in section 6.

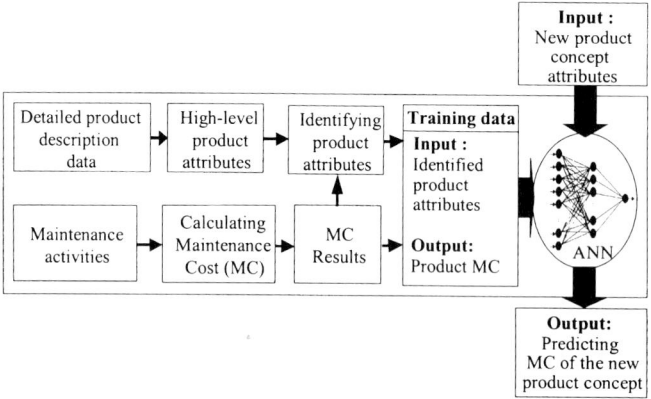

Figure 1. Training process of estimating MC

D002/022 © IMechE 2002

2 BACKGROUNDS AND RELATED WORKD

The conceptual design phase defines the basic characteristics of a product, ranging from cost to environmental impact [2]. Decisions that emerge from the conceptual phase are often locked in due to the large amount of resources: time, manpower, and money, needed to change course as launch deadlines approach. Therefore, it is important that cost considerations are used in the evaluation of concept feasibility along with other requirements. This means designers must be able to evaluate the approximate cost performance of many solution concepts early in the design process.

Time is usually a scarce resource during the product development cycle. Development time can mean the difference between leading or following in an industry; therefore, it limits the ability to create detailed models for many different concepts. Additionally, in conceptual design the lack of information is a significant barrier to the creation of models needed to evaluate different ideas.

Although it is a good idea for product designers to have some knowledge of cost estimation, it is not and should not be their area of primary expertise. Ideally, the services of cost estimators should be extended to designers. Communication, although necessary for such an extension, is often a barrier as it takes time to establish and maintain synchronization of information between designers and cost estimators.

Maintenance can be defined as the combination of all technical and associated administrative actions, including supervision actions, intended to keep an item or system in, or restore it to a state in which it can perform its required function [26]. In other words, maintenance is related to keeping a system's facilities in functioning order. The main purpose of maintenance engineering is to reduce the adverse effects of breakdown and to increase the availability at a lower cost, in order to increase performance and improve the dependability level. Generally, maintenance may be classified as two categories: preventive and corrective maintenance. Preventive maintenance is carried out at predetermined intervals or according to prescribed criteria and is intended to reduce failure probability or functioning degradation of an item. Corrective maintenance is maintenance carried out after breakdown detection and is intended to put an item into a state in which it can perform its required function. In this paper, we focus on corrective maintenance after occurring breakdown of a product or system. Maintainability is adapted as the parameter represented corrective maintenance for a product or system.

Maintainability is an important aspect of life-cycle concerns and plays significant role during the usage phase of a product. It is the design attribute of a product or system which facilitates the performance of various maintenance activities, in particular, inspection, repair, replacement and diagnosis. These activities for a good maintainable product system should not only be performed in quick possible time but also with optimal personnel and support equipment [25]. Moreover, design characteristics of a product or system which facilitate maintainability will be effective if due recognition is given to the factors which support system maintenance during the usage phase. It is indispensable to perceive all the aspects of maintainability right from the design stage of the product in systematic way. This emphasis to develop methodology to evaluate the maintainability of product at design stage qualitatively or quantitatively.

There are some techniques to assess the quality and reliability of products. Failure Modes and Effects Analysis (FMEA) helps design teams to identify, define, and eliminate known or potential failure modes of a product [9, 11]. FMEA focuses on the occurrence, severity, and detection of failures, assigning risk priority estimates to various failure modes. On the other hand, if historical failure rate data is available on the failure modes, the statistical data can be used.

Some studies have been made by researchers in developing procedures for evaluation of the maintenance of various systems. Takata et al. [21] have proposed facility model for life cycle maintenance system. In these models, the authors have evaluated qualitatively the deterioration of the system component arising from operational and environment stresses. This approach is also useful for prediction of failures. Vuiosevic et al. [24] have evaluated the maintainability of the systems on the basis of cost of assembly/disassembly. Balanchard et al. [1] and Cunningham and Cox [5] have evaluated the maintainability of mechanical systems in terms of Mean Time To Repair (MTTR) by considering time taken in disassembly, assembly, localization and isolation of least replaceable component.

Gershenson and Ishii [13] were the first to address serviceability in design. They recognized the drivers of service cost, including part cost, labor cost, and failure rate. This information was incorporated into worksheets that could be applied to analyze any design. Ishii [15] also looks at an integrated life-cycle design methodology that would better allow engineers to understand the life-cycle implications of certain design decisions. Ishii presented specific evaluation methods for calculating service and retirement costs.

Although these existing methods are all useful, they are not ideally suited for early conceptual design. Qualitative information is difficult to use in highly dimensional multi-attribute tradeoffs, and the analytical techniques are still somewhat prohibitive from a modeling viewpoint. This study explores the ANN method to estimate the product MC in conceptual design. In order to estimate MC in conceptual design, the product attributes, maintainability attributes in particular, are identified and the MCs of products are calculated. The identified product attributes as inputs and the calculated MC as output of ANNs are applied. The product designers query this learning model with high-level product attributes to quickly obtain an approximate MC for a new product concept. Designers need to simply provide high-level attributes of new product concepts to gain MC predictions based upon trends inferred from real products and MC studies used as training data.

2 IDENTIFICATION OF PRODUCT ATTRIBUTES

In this section, we identify product attributes as inputs for use in the ANN model. The attributes need to be both logically and statistically linked to MC, and also be readily available during product concept design. The attributes must be sufficient to discriminate between different concepts and be compact so that the demands on the ANN model are reasonable. Finally, they must be easily understood by designers and, as a set. These criteria were used to guide the process of systematically developing a product attribute list.

With these goals in mind, a set of candidate product attributes, based upon the literature and the experience of experts [3-4, 19], was formed. Experts in both product design and cost estimation discussed as candidate attributes derived from the literature. In practice, product attributes at the conceptual stage are defined as a few, simple, and expressed in a product-specific language. Also, different levels of information are available and used at the early stage of product design, depending on the purpose of the design [18]. The candidate product attributes identified initially are listed in table 1[17, 19-20]. They are specified, ranked, binary or not applicable according to their properties such as an appropriate qualitative or quantitative sense or typically rank order concepts.

Table 1. The candidate product attributes

Durability	Selling Price	In use energy source
Strength	Product Liability	In use power consumption
Conductivity	Distribution mass	Modularity
Mass	Distribution volume	Upgradeability
Volume	Transport distance	Serviceability
Materials (Various)	Transportation means	In use flexibility
Performance	Lifetime	Recycled content
Functionality	Use time	Recyclability
Process	Mode of operation	Reusability
Assemblability	Additional consumable	Disassemblability

This study helped us identify attributes that designers could both understand and had knowledge of during the conceptual design. Furthermore, we were able to assess which attributes are likely to vary significantly from concept to concept. As mentioned earlier, corrective maintenance is focused on and maintainability is adapted as the parameter represented the corrective maintenance. Therefore, maintainability attributes are additionally identified.

2.1 Maintainability attributes

Maintainability is one of the system design parameter that has a great impact in terms of ease of maintenance. System failure is inevitable no matter how reliably this is built-in, so its ability to be quickly restored is therefore most important. Maintainability characteristics of the system design facilitates this and leads to lower maintenance man hours, skill level, tools and test equipment with greater system availability. There exists functional relationship between the components and assemblies and these functional relationships and their spatial motion requirements identify the design attributes for the maintenance. The synthesis of structure can change the intricacies of the system for better maintainability. The features are also dictated by maintenance personnel and support variables. It is however necessary to consider these at the design stage, as they influence and are also influenced by the way the product is designed and manufactured. Hence they are also considered as the product attributes at the early design stage.

Maintainability attributes for electronic products, in general, are identified and they are referred as attributes ascribed to the characteristics of the product maintainability. The identified maintainability attributes are presented in table 2 [16, 22-23, 24-25]. They are also estimated by an appropriate qualitative or quantitative sense as shown in table 3.

Table 2. Identified maintainability attribute for products

Accessibility	Standardization	Diagnosability
Interchangeability	Simplicity	Tribo-concepts
Reasemblability	Redundancy	Ergonomics
Modularity	Identification	

Table 3. Qualitative or quantitative criteria for maintainability attribute – Simplicity

No.	Description of scoring criteria	Score
1	Minimum number of components and assemblies are used find design is not complex	4
2	Design fulfils one of earlier condition	2
3	Design does not fulfil any of the above requirement	0

Sampling data with product attributes and corresponding MC from actual historic maintenance activities were collected for different electronic products. Based upon these data, the candidate attribute set was again refined and then tested for first order relationships with the MC. Bivariate correlations were be computed and correlation tests to 95% statistical significance were performed between quantitative attributes and the data of MC for various products.

Table 4. An example of correlation coefficients: product attributes vs. MC

Product attributes	The coefficient of correlation
mass	0.49
lifetime	0.43
usetime	0.61
operation mode	0.61
energy source	0.05
power consumption	0.02
:	:
:	:
flexibility	-0.02
upgradability	0.47
modularity	0.51
accessibility	0.56
reassemblability	0.62
simplicity	0.47

The product attributes strongly correlated with the MC are used to predict the product MC. Finally, 24 product attributes for MC are chosen as shown in table 5 and used as inputs in ANN models.

D002/022 © IMechE 2002

Table 5. The final list of identified product attributes for MC

Q – Mass (kg)	Q – Other materials (%mass)	Q – Modularity
Q – Ceramics (%mass)	B –Assemblability	Q – Accessibility
Q – Fibers (%mass)	B – Disassemblability	Q – Reassemblability
Q – Ferrous metals (%mass)	Q – Lifetime (hours)	Q – Standardization
Q – Non-ferrous metals (%mass)	Q – Use time (hours)	Q – Simplicity
Q – Plastics (%mass)	D – Mode of operation	Q – Identification
Q – Paper/Cardboard (%mass)	B – Serviceability	Q – Diagnosability
Q – Chemicals (%mass)	B – Upgradeability	(* Q: Qualitative or Quantitative,
Q – Wood (%mass)	Q – Modularity	D: Dimensionless, B: Binary)

4 CALCUATING MAINTENANCE COST

In this section, the procedure of calculating MC of products is described. Considering life-cycle constraints can be important when making design decisions for products that require regular maintenance. In such situations, product design need to address maintenance concerns, specifically the trade-offs involved between reliability, part cost, and others. Given the nature of these requirements, the MC for a given product should be understood prior to making design decisions of products. In order to calculate the MC of products, the parts of previous works [7, 13, 15] are applied to our approach. Adapting the modified form of their equations, the MC (corrective) can be computed.

The MC of usage phase was calculated by the following equation. Equation (1) is used to take into account labour cost, part cost, and failure rate.

$$C_{Maintenance} = [(LC_{Fixed} + (L_T \times L_R) + C_R)] \times F_R \qquad (1)$$

Where:

$C_{Maintenance}$ = MC (corrective) ($)
LC_{Fixed} = Fixed labour cost ($)
L_T = Labour time (h)
L_R = Labour rate ($/h)
C_R = Replacement cost of parts or materials ($)
F_R = Failure rate

In equation (1), LC_{Fixed} refers to fixed labour cost when a maintenance representative visits a customer site. The labour time, L_T, is the maintenance time such as Mean Actual Repair Time (MART) or Mean Actual Maintenance Time (MAMT) associated with repairing or replacing the individual components that fulfil the primary function. The cost of the replaced parts or materials is the mean replacement cost of products and represented by the value C_R. The failure rate, F_R, is obtained by the reliability information of products. In this paper, we use constant and independent failure rates which is appropriate for our earlier conceptual design phase.

The above MC equation only reflects the costs borne by the company that produces the product. In short, the equations provide an internal view of maintenance. The equations do not reflect the cost or hardship experienced by the customer. Therefore, the company needs to examine more maintenance policy from the customer's point of view.

In selecting suitable product design, designers should know the life cycle concerns such as product maintenance based on the labour cost, part costs and reliability information. Using these information, effective product designs can be defined and selected. By taking into these considerations, designers can better understand how MCs are affected by product design. In turn, they can be sure to incorporate such concerns into product design decisions.

Table 6. Examples of the calculating the MC of products

Product	Failure part	Fixed labor cost($)	Labor time (h)	Labor rate ($/h)	Mean replacement cost of parts($)	Failure rate (%)	Life time (yrs)	Mainternace cost($)
Washing Machine	Gear mechanism Magnet of water supply M-PCB Gear spring Water valve, etc.	$10	0.5	$20	26.03	4.87	10	$22
Refrigerator	Frost timer M-PCB Louver assembly Freezer Compressor Control S/W, etc.	$10	0.5	$20	30.61	3.23	10	$16

5 EXPERIMENTS AND DISCUSSION

With product attributes and MC, ANN-based learning models were trained in an effort to validate the proposed method. Sampling data with product attributes and corresponding MCs were collected for 40 different electronic products. The examples of sampling data for the ANN model are shown in table 7.

Table 7. Examples of sampling data for the ANN model

Product	Inputs								Output
	Mass (kg)	Ferrous M. (%mass)	Plastics (%mass)	Lifetime (hours)	Use time(hrs) (hours)	...	Power consump. (watt)	Modularity (0–4)	Maintenance cost($)*
1	8.17	32.62	61.58	61320	3041	...	1064	1	6.00
2	1.04	16.19	77.65	26280	13	...	58	1	1.79
3	0.18	45.77	32.86	43800	13688	...	0	1	8.25
4	0.64	22.16	71.09	2160	45	...	13	1	1.85
5	1.93	2.85	65.54	43800	487	...	616.44	4	2.75
...	...	...	...	...	...	...	...	...	...
...	...	...	...	...	...	...	...	...	...
38	49.78	67.07	27.64	87600	87600	...	13	3	18.00
39	40.46	8.83	25.81	87600	11680	...	616	3	6.54
40	35.01	24.24	51.75	121764	121764	...	19	4	12.57

The multiple-layer neural network with back propagation (BP) training [12, 14] was used to estimate the MC of products. In order to decide the structure of BP neural network, the convergence rate of error was checked by changing the number of hidden layers, the number of nodes in each layer and by adjusting learning rate η, and momentum term α. Here, η and α are constants whose values are between 0 and 1. More than 50 experiments were performed to determine the best combination of the learning rates (η), momentum term (α), number of hidden layers, number of neurons in hidden layers, learning rules and transfer functions. The resulting network possessed a hidden layer with 20 neurons. The most popular learning rules, generalized delta rules and a sigmoid transfer function were used for the output node. Figure 2 shows the structure of the back propagation neural network, which consists of an input layer with 24 nodes, a hidden layer with 20 nodes and an output layer with one node.

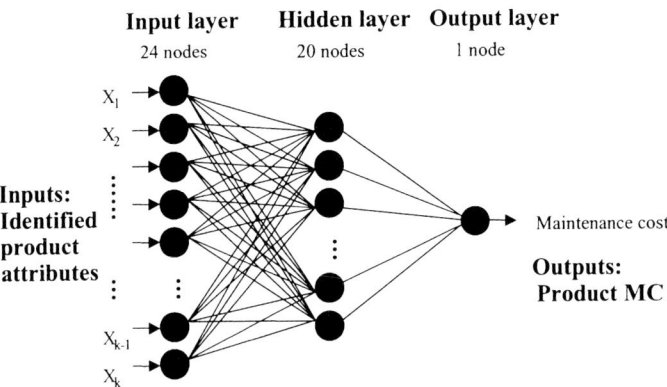

Figure 2. Structure of the backpropagation neural network to estimate the product MC

The BP neural network to estimate MC of products was implemented in C++. The training of the back propagation neural network took 102 seconds for 30 learning patterns on a 500 MHz Pentium III processor. When η and α were 0.6 and 0.25 respectively, the number of iteration was 5,000, and the mean square error was 0.006167.

In order to test the capability of the trained ANN, ten another samples are used as the test set which have not been used in the training. The results of the product LCC predicted by the ANN model for ten products are provided in table 8.

Table 8. The predicted results of product LCC by ANN

Product	Actual MC ($)	Predicting MC ($)	Relative error (%)
1. Vacuum Cleaner	6	6.02	-0.38
2. Mini Vacuum	1.79	1.75	2.08
3. Radio	9.46	9.44	0.12
4. Heater	3.23	3.22	0.15
5. Coffee Maker	2.83	2.81	0.97
6. Washing Machine	22	22.01	-0.02
7. Refrigerator (S)	12.38	12.38	0
8. Refrigerator (L)	16	16.01	-0.02
9. TV	6.54	6.68	-0.06
10. LCD TV	12.57	12.59	-2.11
Ave. absolute error			0.59
Max. absolute error			2.11

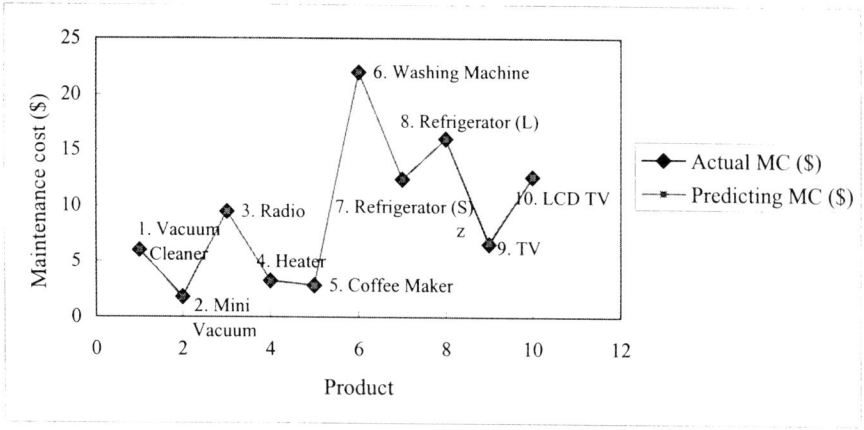

Figure 3. Comparison results of the MC of products

In table 8, some observations from the testing results are summarized as follows: (1) Among ten testing samples, it can be found that the percentage errors of the samples are within ±3%. This is considered as a very good MC estimation at the early design stage. (2) The maximum deviation of MC estimate from the actual MC is about 2.11% which is acceptable for practical use. (3) The performance of the trained ANN is consistent in the training, validation and testing samples.

During the early conceptual design stages of product development, available data are limited and the cost analyst must depend primarily on the use of various parametric cost estimating techniques in the development of cost data. The accuracy of a cost model deviated from an actual cost in the conceptual design phase is typically between –30 and +50% [6], so the proposed method based on ANN shows better product MC estimation and gives the guidelines for the cost-effective design decisions in conceptual design phase.

The two advantages of the approximate MC method are summarized: (1) The product attributes include the cost aspects related to products. Extracting such product attributes can be easily done by a product designer. Detailed information for cost estimation is not required. (2) The ANN based method can help better conceptual design through evaluating the costs of different design alternatives.

6 CONCLUSIONS

It has been recognized that the design process needs cost models that: (1) take into account the life cycle of products, (2) can be used at the very early stages of design and (3) can provide information to designers in a timely manner and in a form that can be understood and used. Some efforts have been made toward providing the designer with cost information during the design process. The product LCC mainly determined by early design decisions. But, at the early conceptual design stage designers do not know the costs incurred in subsequent life

D002/022 © IMechE 2002

cycle phases. Thus, the estimation method for the product LCC should be able to offer sufficient cost estimation in response to design decisions and design guidelines.

The lack of analytic cost estimation for early conceptual design motivated the development of the estimation method. This paper described procedures to develop a foundation for estimating the product MC as LCC in conceptual product design. Three areas critical to the preliminary validation of the approaches were explored: model output of product MC; model inputs in the form of a compact, meaningful, and understandable set of concept attributes; and the capability to estimate the product MC using the ANN model.

It is apparent that the proposed method using ANNs is feasible for estimating the MC at early conceptual design stages. The proposed method can be used to estimate the product LCC and gives the guidelines leading to cost-effective design decisions in the conceptual design phase. . Particularly, it can: (1) help in designing decisions by means of identification of product attributes which affect a maintenance, (2) estimate the quantitative MC of various products, (3) identify any changes in design or the requirement or both, which are necessary to meet maintenance requirements within the given constraints.

In future, the various product attributes for LCC factors besides MC are identified and further tests using more data are needed to determine to what extent the ANN model can provide reasonable estimations for product attributes and to test for another LCC factors.

REFERENCES

1 **Balanchard, BS., Verma, D., Peterson, EL.,** (1995) Maintainability, New York: Wiley and Son.
2 **Bhamra, T. A., Evans,S., McAloone, T. C., Simon, M., Poole, S.** and **Sweatman, A.,** (1999) Integrating Environmental Decisions into the Product Development Process: Part 1 The Early Stages, Proceedings of the First International Symposium on Environmentally Conscious Design and Inverse Manufacturing, Tokyo, Japan, February 1999, Los Alamitos, CA: IEEE (Institute of Electrical and Electronic Engineers), pp.329-333.
3 **Brezet, H.** and **van Hemel, C.,** (1997) Ecodesign, a Promising Approach to Sustainable Production and Consumption. Paris, France: United Nations Environmental Program (UNEP) Industry and Environment, 1997.
4 **Clark, T.** and **Charter, M.,** (1999) Eco-design Checklists for Electronic Manufacturers, Systems Integrators, and Suppliers, of Components and Sub-assemblies. http://www.cfsd.org.uk.
5 **Cunningham, CE., Cox, W.,** (1972) Applied maintainability Engineering. New York: Wiley and Son.
6 **Creese, R. C.** and **Moore, L. T.,** (1990) Cost modelling for concurrent engineering, Cost Engineering, Vol.32(6), pp. 23-27.
7 **Dahmus, J. B., Otto, K. N.,** (2001) Incorporating lifecycle costs into product architecture decisions, ASME Design Engineering Technical Conferences and Computers and Information in Engineering Conference Pittsburgh, Pennsylvania, September 9-12.
8 **Dekker, R.,** (1996) Application of Maintenance Optimisation Models: A Review and Analysis, Journal on Reliability Engineering of System Safety, Vol. 51 (3), pp. 229–240.

9 **Eubanks, C., Kmenta, S.** and **ishii, K.,** (1997) Advanced Failure Modes and Effects Analysis Using Behaviour Modelling ASME Design Engineering Technical Conferences, Sacramento, California. DTM-3872.

10 **Fabrycky, W. J.** and **Blanchard, W. J.,** (1991) Life cycle Cost and Economic Analysis, Englewood Cliffs, NJ: Prentice Hall.

11 **FMEA,** (1995) Potential Failure Modes and Effect Analysis, Reference Manual, Second Edition, SAEJ-1739 Equivalent, Automobile Industry Action Group.

12 **Freeman, J. A.** and **Skapura, D. M.,** (1992) Neural networks: algorithm, application and programming techniques, Addison-Wesley Pub Co.

13 **Gershenson, J.** and **Ishii, K.,** (1991) Life-Cycle Serviceability Design, ASME Design Engineering Technical Conferences, Miami, Florida. DTM, pp. 127-134.

14 **Hassoun, M. H.,** (1995) Fundamentals of Artificial Neural Networks (Cambridge, MA: MIT Press).

15 **Ishii, K.,** (1995) Life-Cycle Engineering Design, Journal of Mechanical Design, Vol. 117: 42-47.

16 **Paasch, RK., Ruff, DN.,** (1997) Evaluation of failure diagnosis in concepts design of Mechanical systems, Trans. of ASME: J. of Mech. Design, Vol.119 (1), pp.57-67.

17 **Park, J.-H., Seo, K.-K.,** and **Wallace, D.,** (2001) Approximate Life Cycle Assessment of Classified Products using Artificial Neural Network and Statistical Analysis in Conceptual Product Design, Proceedings of the Second International Symposium on Environmentally Conscious Design and Inverse Manufacturing, Tokyo, Japan, February 2001. Los Alamitos, CA: IEEE (Institute of Electrical and Electronic Engineers), pp. 321-326.

18 **Potts, A.,** (2000) Personal communication. Principal Designer, Potts Design, Stoneham, MA.

19 **Rombouts, J P.,** (1998) LEADS-II. A Knowledge-based System for Ranking DfE-Options, Proceedings of the 1998 IEEE International Symposium on Electronics and the Environment. IEEE (Institute of Electrical and Electronic Engineers), pp.287-291.

20 **Sousa, I., Eisenhard, J. L.,** and **Wallace, D.,** (2001) Approximate Life-Cycle Assessment of product Concepts Using Learning Systems, Journal of industrial Ecology, Vol .4 (4): 61-81, 2001.

21 **Takata, S., Hiraoka, H., Asama, H., Yamoka, N., Saito, D.,** (1995) Facility model for lifecycle maintenance system, Annals of the CIRP; Vol.44 (1), pp.117-121.

22 **Tarelko, W.,** (1995) Control model of maintainability level, Reliability Engineering and System Safety; Vol. 47 (2), pp. 85-91.

23 **Utez, H.,** (1983) Maintainability of production system, Maintenance Management International, Vol. 4, pp. 55-68.

24 **Vujosevic, R., Raskar, R., Yeturkuri, NV., Jothishankar, MC., Juang, S-H.,** (1995) Simulation, animation and analysis of design disassembly for maintainability analysis, Intl. J. of Production Research, Vol. 33 (11), pp.2999- 3022.

25 **M.F. Wani, O.P. Gandhi,** (1999) Development of maintainability index for mechanical systems, Reliability Engineering and System Safety, Vol.65 (3), pp. 259-270.

26 **Simeu-Abazi Z., Sassine, C.,** (2001) Maintenance Integration in Manufacturing Systems: From the Modeling Tool to Evaluation, The International Journal of Flexible Manufacturing Systems, Vol.13 (3), pp. 267–285.

A methodology to support the implementation of product recovery

A RAHIMIFARD, S T NEWMAN, and **S RAHIMIFARD**
Wolfson School of Mechanical and Manufacturing Engineering, Loughborough University, UK

ABSTRACT

Product Recovery (PR) and End-of-Life management (EoL) are among the main research areas related to design and manufacture for sustainable development. These concepts are developed to support the transformation of used and discarded products into useful condition through re-manufacture, re-use and recycling. The main motivations are economic gains and complying with national and international legislation, but also include the reduction of negative impacts on environment and consequently improving the public image. The research reported in this paper has developed a five-stage methodology to support the EoL management within manufacturing companies. The application of this methodology has been demonstrated through the generation of a systematical framework for the recovery of the cutting tools at the end of their useful life, both for re-use within a company and across the external tool supply chain.

1. INTRODUCTION

Increased public awareness towards the global environmental problems has forced manufacturing companies to consider the negative impacts of their activities on environment. In this context, the effective management of products at the end of their useful life will become more crucial in the future due to the increasing amount of national and international legislation and directives aimed at making take-back and recovery of used products obligatory for the Original Equipment Manufacturer (OEM). This has highlighted a need for a systematic approach for enhancement of information, business and production management systems to deal with additional activities and processes related to the recovery of products.

The effective PR and EoL management of products within any manufacturing enterprise requires a major alteration on both the internal and external business and operational structures. Clearly, there should be an in-depth understanding of such implications on the existing manufacturing activities before considering any major modification of business and operational processes. It is crucial to assess and evaluate alternative approaches in product take-back and recovery before final decisions are made.

The research reported in this paper has generated a novel systematic five-stage methodology to support EoL management. The initial part of the paper provides a review of relevant literature, together with a description of the various stages involved in this methodology. The later sections discuss the application of this novel approach for the recovery of the cutting tools within machining applications and analyses the implications of modification to the existing manufacturing activities to include recovery and recycling of cutting tools.

2.   REVIEW OF RELEVANT RESEARCH

The increasing significance of product recovery within manufacturing activities has brought a corresponding influence in the research covering the production life cycle stages from product design to final disposal. In the design stage, new concepts have emerged such as design for disassembly, re-manufacturing, re-use, and recycling, which incorporate end-of-life decision considerations as design objectives (1 , 2). Within the production stage, operational management issues related to product recovery include disassembly and re-manufacturing planning and control (3, 4, 5). The important issue is to find a balance between the cost of disassembly and re-manufacturing and the returned benefits, as explored by (6) and (7). There are also a number of studies exploring the use and suitability of a Material Requirement Planning (MRP) based approach with some modifications for scheduling in recovery environments (3 , 8). Other researchers propose and investigate the use of alternative approaches for scheduling and control at the shop floor such as the drum-buffer-rope concept (9), flexible KANBAN (10), Push and Pull control strategies (11). Other issues investigated in this area relate to inventory control requirements within recovery systems which differ from traditional manufacturing systems due to a high degree of uncertainty in timing, quantity and quality of returned products and the demand for recovered parts (12, 13, 14).

One of the critical decision making tasks related to end-of-life (EOL) management is the selection among the options of the recovery of the used product as a whole, component or part recovery, material recovery (recycling) or disposal. The objective is to maintain the profitability and not to violate the technical feasibility constraints (15, 16, 17, 18, 19). Finally, a significant body of research has explored the new material flows within PR applications from the user to the producer, which include the collection and transportation processes involved in what is referred to as reverse distribution (18 , 20).

3.   A   METHODOLOGY   TO   SUPPORT   THE   IMPLEMENTATION   OF PRODUCT RECOVERY

The EoL management concepts extend the traditional view of the manufacturing supply chain to include the activities, actors and the structures required to accomplish recovery at the end of products' life. This has resulted in the emergence of the new supply chain concept, referred to as 'product recovery supply chain'. The additional activities in the product recovery supply chain as opposed to the traditional manufacturing supply chain are the collection, assessing, sorting, re-processing and redistribution of the used products and the disposal of waste. The recovery of used products can be classified into a number of levels as outlined below :-

* *Product recovery* is the reintroduction of the used product back into the market through a series of processes such as inspection, disassembly, replacing or repairing bad components and re-assembling (often referred to as re-manufacturing).
* *Module & part recovery* where a subset of parts and components of used products can be recovered, repaired or re-conditioned for re-use in production of new products.
* *Material recovery* is retrieving the material content of the whole or a subset of the components of used products through a range of processes at the end of which the identity of the product is completely lost (often referred to as recycling), and finally,

- *Energy recovery* where in limited applications some of the material not recovered through one of aforementioned processes is used to generate energy (often in the form of heat and electricity).

It should be noted that in most applications due to the economical and environmental implications such as cost, effort, time, and energy associated with re-production, the most preferred approach to recovery in the first place should be directed at the recovery of the product, followed by modules and parts. The latter two options, namely material and energy recovery should be considered in industries that are characterised by short technology cycles and high technological obsolescence. The research reported in this paper has generated a novel systematic five-stage methodology, referred to as the 'Product Recovery Implementation MEthodology' (PRIME), as outlined below :-

i)   *Technological Assessment*: the evaluation of available technologies in various applications to identify the most appropriate PR procedures for a specific product, resulting in adoption of one or a combination of re-manufacturing, re-use or recycling approaches.

ii) *Business and Economical Evaluation*: this includes a cost-benefit analysis, investigation of the relevant national and international legislation, identification of the marketing implications for both the original and recovered products, and business process planning to include new processes and actors.

iii) *Resource Assessment*: the evaluation of various required resources including the internal and external hardware, software and human resources.

iv) *Logistics and Operation Planning*: Product take-back logistics, planning and control of recovery processes, inventory control of the new and recovered products.

v)   *Information Specification*: identification of product and manufacturing information related to additional activities included in a product recovery supply chain.

There are basically two possible business solutions for the realisation of product recovery within manufacturing applications, namely (a) through the addition of recovery capabilities to the Original Equipment Manufacturer (OEM) activities or (b) by third party independent recovery companies whose sole business is to re-process used products. This correspondingly highlights two possible reference configurations for the realisation of PR procedures within a manufacturing enterprise, namely:-

- *Reference Configuration 1 - Recovery by Manufacturer :* in this reference configuration the original manufacturer of the product takes the products back at the end of their life and carries out the recovery processes in-house, as illustrated in Figure 1a. This is often achieved by the expansion of the business and manufacturing facilities to include the required resources to undertake the recovery processes. The adoption of this reference configuration provides more control over the secondary market for the OEM, and the information required for disassembly and remanufacturing is readily available within the company which makes the implementation of product recovery procedures easier.

- *Reference Configuration 2 – Recovery by Independent Recoverer* : in the second reference configuration, an independent recovery company undertakes PR on behalf of one or more OEMs (see Figure 1b). The recoverer receives the products from the collectors at the end of their life and carries out the required recovery processes. Then, the recovered products are either supplied back to the original manufacturers or sold on to secondary customers. Clearly, in this reference configuration a particular OEM has less control over the secondary market, and issues related to product confidentiality and EOL information management is much more complex. The independent recoverer often has access to a vast amount of information supplied by a number of OEMs, and has the additional advantage of supplying the recovered products to a much larger market.

There are a large number of factors influencing the suitability of one of these reference configurations for a particular manufacturing company, including product size and type, process complexity, production capacity, geography of the initial distribution, and relevant legislation. These factors should be carefully analysed and assessed before establishing the suitability of one of these recovery reference configurations. For example, within a company where manufacturing activities are mainly consists of assembly of large and specially designed products in small batches, the first reference configuration is considered to be more attractive than in the case of a high volume components manufacturing company. One of the other key decision issues is the geographical distribution of customers and effort involved in the collection of used products which may make the first reference configuration infeasible. Furthermore, the PR processes such as disassembly, repair and reassembly will require additional planning, control and co-ordination which could result in undesirable complexity and higher production cost. The second reference configuration provides OEMs with the advantage of outsourcing the recovery processes thus not needing a major alteration in their production facilities. Such a shared recovery approach may significantly reduce the cost of recovery processes. However, due to the complexity of recovery technologies and unfamiliarity of such re-manufacturing business concepts, in many industrial sectors these independent recoverers may not exist at present.

The remaining sections of this paper illustrate the application of PRIME and the advantages and disadvantages associated to aforementioned reference configurations for the recovery of cutting tools within machining applications.

4.   THE APPLICATION OF PRIME FOR RECOVERY OF CUTTING TOOLS

Numerous innovations in cutting tools design and manufacturing have appeared to improve productivity and quality, using new cutting materials and more complex geometry. Cutting tools can be categorised based on various types of machining operation such as milling, drilling, turning, grinding, boring, and tapping.

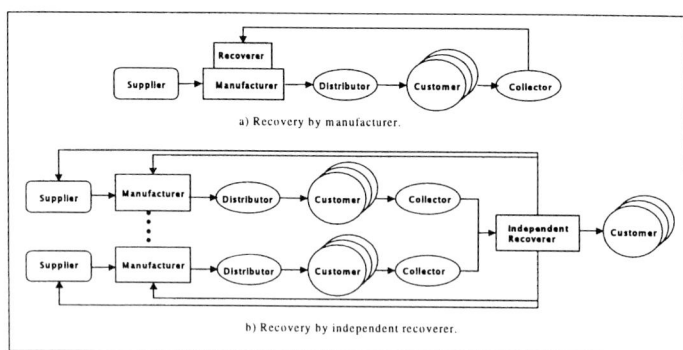

Figure 1 : Reference configurations for product recovery realisation

                                                                                    D002/010 © IMechE 2002

These tools often consist of an assembly of 3-4 components such as tool holders, shanks, collets, tool extensions, cutting tips or edges and gradually lose their accuracy in cutting capabilities over a period, in which case they are referred to as worn or spent. The period in which a tool can be used for the machining operation with certain accuracy is referred to as 'tool life' which is defined in terms of a number of minutes or hours of cutting time or the number of parts manufactured. In machining applications, the cost of cutting tools and their timely provision is often as much as 15% to 40 % of overall production costs. Cutting tool management is a very complex subject as it is influenced by a wide range of different activities within the production system, and plays a crucial role for the smooth running of a machining facility. The management of cutting tools consists of a range of activities including purchasing, storage, inventory control, assembly, pre-setting, kitting, tool requirement planning, and scheduling of transportation between stores. The planning for tool flow can be considered, using a hierarchical division of tool stores and tool buffers. There are three common storage areas for cutting tools within machining facilities, as illustrated in figure 1. These are :-

- *central tool store/room* where all the components of cutting tool assemblies are stored. The activities of tool assembly, disassembly, rework and presetting take place within the central tool store/room.
- *cell tool store* is a buffer between the central tool store and machine tool magazine. The assembled, preset and new tools, required in the short term (e.g. a shift, a day) within the machining cell are temporarily stored in the cell tool store. Similarly, the worn tools, which are recently taken off the machines, are temporarily held in the cell tool store, before being transported back.
- *machine tool store* which is often in the form of a tool magazine (e.g. tool chain, drums or cylindrical magazine) where a set/kit of cutting tools are held, corresponding to operation sequences of the assigned parts.

The planning for the cutting tool flow is often based on a tool requirements planning (TRP) which is the task of calculating the net tooling requirements over a time period (e.g. a shift, a day, a week) and the generation of appropriate transportation schedules between the central and the cell tool stores, and eventually from the cell tool store to the machine tool magazines.

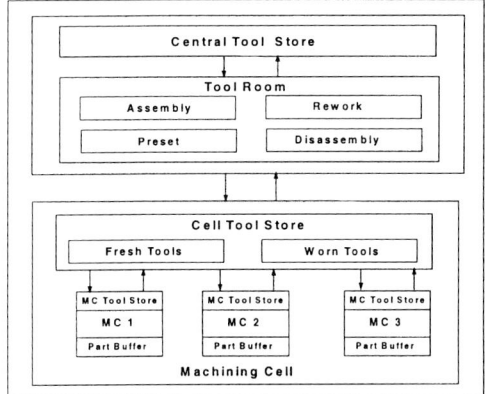

Figure 2 : A typical tool flow configuration

The major assertion made in this paper is that the adoption of a recovery or recycling procedure for cutting tools promises to provide significant financial benefits within majority of machining applications. Tool recovery or recycling in this research is defined as the ability to regrind or rework a worn tool as either (a) reground tool, (b) reincarnation of the tool to another tool or (c) material reclamations. Furthermore, as stated previously cutting tools are often consisting of three parts, namely shanks, collets which are counted as durable, and cutting tips as consumable. Clearly, in the first instant the main emphasis on the tool recovery/recycling should be directed at the consumable parts, i.e. the cutting tips. The recovery and recycling of the cutting tools corresponds to the part/module recovery and material recovery defined in the four-level PR classification outlined in section 3. There is a framework of three possible solutions for the realisation of cutting tool recovery within machining applications, namely :-

i)   through addition of regrinding/rework capabilities within a user company
ii)   by returning the tool the original tool supplier
iii)   through a third party independent recovery/recycling companies whose sole business is to re-process used tools.

In this framework, options 1 and 2 are based on the first reference configuration defined earlier in the paper, whereas option 3 represents the second reference configuration. Clearly the implementation of one or more of these recovery options should be investigated using a systematic approach as outlined by PRIME and discussed below :-

•   *Technological assessment* : there are several materials used for making cutting tools of which the most commonly used are high speed steel (HSS), ceramics, and carbide . Clearly, a different range of processes and resources are required to recover/recycle cutting tips/edges made from each of these materials. In addition, the geometrical shapes and sizes of cutting tools significantly influence the decision for tool regrinding, tool reincarnation or material reclamation.

•   *Business and Economical Evaluation* : Clearly, the type of raw material being machined in an application (e.g. aluminium vs. steel) has a significant influence in the selection of one of the tool recovery options and their economical viability.

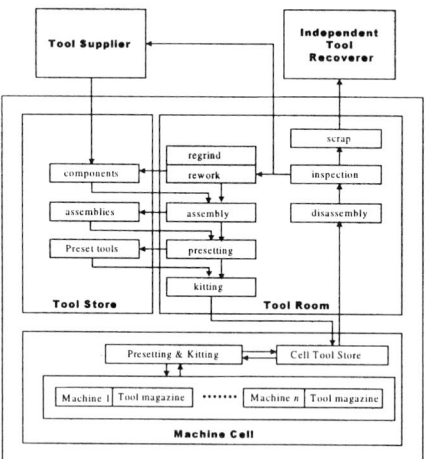

Figure 3 : A framework for recovery/recycling of cutting tools

   D002/010 © IMechE 2002

In addition, from the tool recoverer's point of view, there is a potential to re-sale the cutting tools initially used in a high precision application (e.g. aerospace) to a mass volume production industrial sector (e.g. automotive).

- *Resource Assessment* : the issue of resource requirements for tool recovery/recycling has attracted significant attention in the recent years, resulting in the reduction of the cost of reground cutters and consequent increase in their use within machining facilities. Furthermore, there is an ever increasing willingness among the tool suppliers for taking back their products at end of their life and also a rapid rise in the number of independent tool recoverer/recyclers.

- *Logistics and Operation Planning* : the geographical position of tool manufacturers, distributors, and recyclers has a significant influence on the selection of the recovery option, as this determines the logistics complexities and delays involved. In addition, there is a requirement for the enhancement of TRP to include systematic approach for the return of the used tools to a central storage for assessment and routing within one of the tool recovery options.

- *Information Specification* : the additional information required to support the tool recovery includes the data related to the jobs and the machines in which the tools are used, the number of parts and material machined by the tool, the remaining tool life, possible recovery options for the tool, the original tool supplier, etc.

5. CONCLUDING DISCUSSION

The significant attention paid to the environmental impact of manufacturing activities in recent years, highlights the paramount importance and inevitability of the inclusion of product recovery procedures within an increasing number of manufacturing applications in the near future. At present the recovery of products may not be economically viable in many industrial sectors. This has resulted in a lack of significant and a consistent desire by the manufacturing companies and in particular the small to medium enterprises (SMEs) to adopt the PR procedures. The authors argue that the required and desired levels of reduction in the amount of negative impacts of manufacturing activities on the environment can only be achieved through development of simple and economical approaches for the adoption of PR within SMEs.

The review of literature has shown that the previous PR research work has concentrated on isolated topics rather than developing a holistic view linking and integrating all aspects of product recovery procedures. The research reported in this paper has defined a systematical methodology for the implementation of PR procedures. The future research will aim to generate a CASE software tool to support the various stages of the PRIME methodology for each reference configuration. This CASE tool will include a knowledge based advisory system for PR legislations and technologies, a business model, a planning model, supported by the product and resource information models which are enhanced to include the PR related information.

In majority of machining applications one of the most economically viable recovery procedures relates to the recovery and recycling of the cutting tools, as tools represent a significant proportion of the total production cost. The paper has illustrated the application of PRIME to recovery/recycling of the cutting tools by definition of a framework of three possible options. It is intended to utilise the example of the realisation of cutting tool recovery to test and validate the functionality of the CASE software tool.

6. REFERENCES

1. **Alting, L., Legarth, J.B.,** (1995) Life cycle engineering and design, Annals CIRP, 44 (2), 569-580.
2. **Harjula, T., Rapoza, B., Knight, W.A., Boothroyd, G.,** (1996) Design for disassembly and the environment, Annals CIRP, 45 (1), 109-114.
3. **Gupta, S.M., Taleb, K.N.,** (1994) Scheduling disassembly, International Journal of Production Research, 32 (8), 1857-1866.
4. **Gungor, A., Gupta, S.M.,** (1999) Issues in environmentally conscious manufacturing and product recovery: a survey, Computers and Industrial Engineering, 36(4), 811-853.
5. **Lambert, A.J., Jansen M.H., Splinter M.A.,** (2000) Environmental information systems based on enterprise resource planning, Integrated Manufacturing Systems, 11(2), 105-112.
6. **Lambert, A.J.,** (1997) Optimal disassembly of complex products, International Journal of Production Research, 35(9), 2509-2523.
7. **Navin-Chandra, D.,** (1994) The recovery problem in product design, Journal of Engineering Design, 5(1), 65-86.
8. **Guide, V.D.R., Kraus, M.E., Srivastava, R.,** (1997) Scheduling policies for remanufacturing, International Journal of Production Economics, 48 (2), 187-204.
9. **Guide, V.D.R.,** (1996) Scheduling using drum-buffer-rope in a remanufacturing environment, International Journal of Production Research, 34 (4), 1081-1091.
10. **Kizilkaya, E., Gupta, S.M.,** (1998) Material flow control and scheduling in a disassembly environment, Computers and Industrial Engineering, 35 (1-2), 93-96.
11. **Van der Laan, E., Salomon, M., Dekker, R.,** (1999) An investigation of lead-time effects in manufacturing/remanufacturing systems under simple PUSH and PULL control strategies, European Journal of Operational Research, 115 (1), 195-214.
12. **Fleischmann, M., Boemhof-Ruwaard, J.M., Dekker, R., van der Laan, E., van Nunen, J.A.E.E., van Vasenhove, L.N.,** (1997) Quantitive models for reverse logistics: a review, European Journal of Operational Research, 103, 1-17.
13. **Guide, V.D.R., Jayaraman, V., Srivastava, R.,** (1999) Production planning and control for remanufacturing: a state-of-art survey, Robotics and Computer Integrated Manufacturing, 15(3), 221-230.
14. **Richter, K., Sombrutzki, M.,** (2000) Remanufacturing planning for the reverse Wagner/Whitin models, European Journal of Operational Research, 121 (2), 304-315.
15. **Johnson, M.R., Wang, M.H.,** (1998) Economical evaluation of disassembly operations for recycling, remanufacturing and reuse, International Journal of Production Research, 36 (12), 3227-3252.
16. **Krikke, H.R., van Harten, A., Schuur, P.C.,** 1998, On a medium term product recovery and disposal strategy for durable assembly products, International Journal of Production Research, 36 (1), 111-139.
17. **Low, M.K., Williams, D.J., Dixon, C.,** (1998) Manufacturing products with end-of-life considerations: an economic assessment to the routes of revenue generation from mature products, IEEE Transactions on Components, Packaging and Manufacturing Technology-C, 21 (1), 4-10.
18. **Jung, L.B., Bartel, T.J.,** (1999) Computer take-back and recycling: an economic analysis for used consumer equipment, Journal of Electronics Manufacturing, 9 (1), 67-77.
19. **Goggin, K., Browne, J.,** (2000) The resource recovery level decision for end-of-life products, Production Planning and Control, 11 (7), 628-640.
20. **Klausner, M., Hendrickson, C.T.,** (2000) Reverse-logistics strategy for product take-back, Interfaces, 30 (3), 156-165.

Methods and tools for the development of environmentally sound products

S LEIBRECHT and **R ANDERL**
Department of Computer Integrated Design, University of Technology-Darmstadt, Germany

ABSTRACT

The development of environmentally sound products requires the consideration of processes and the emissions caused by the products throughout their entire life. Emissions must be determined during product design by simulating the whole product life in this stage. Methods and tools for this purpose, which are flexible and adaptable to industrial requirements, are developed within the Collaborative Research Center 392 (SFB392) at the Technical University in Darmstadt.

This paper discusses requirements and approaches for the implementation for an integrated development tool for environmentally sound products. The topics include the motivation and need for such tools, requirements that are important for the application in industry and the attempt for a realization used in the SFB392. A prototype of such an integrated development tool based on object-oriented information modeling and CORBA architecture has been implemented within the SFB392. It is described in chapter 4. The content of this paper reflects the main points of the analysis stage for the next version of the product development environment.

1. INTRODUCTION

The preservation of our environment is a very important topic, discussed often and on from different views and as a result of different circumstances. The answer to the question "why" we need to preserve our environment is clear enough for most people. Nevertheless, a short motivation for the preservation of our environment and the development of environmentally sound products is given in chapter 1.1. The basic idea of an environmentally sound product is explained in chapter 1.2.

1.1 Motivation

Why do we need environmentally sound products? The ecological footprint is one way to measure the human impact on nature. The unit for the ecological footprint is the area of productive land and water a person needs for the resources he/she consumes. This number mainly depends on the way and standard of living, and differs immensely for different countries. The ecological footprint is just a rough estimation, like every method that brings down a huge complexity to just a single number. Nevertheless it shows the tendencies clear enough. Some examples for the average ecological footprint per habitant of different countries are (source: [10], numbers from 1996):

- *USA:*   *12,22 ha*
- *Germany:*   *6,31 ha*
- *Brazil:*   *2,60 ha*
- *India:*   *1,06 ha*

The world wide average is 2,85 ha per person. To be in ecological balance, the following simple equation has to be fulfilled:

```
               [productive area / world]

                          ≥

   [ecological footprint / inhabitant] * [number of inhabitants / world]
```

As long as this condition is fulfilled, the supply of the needed resources is secured. Unfortunately, the numbers are not very promising:

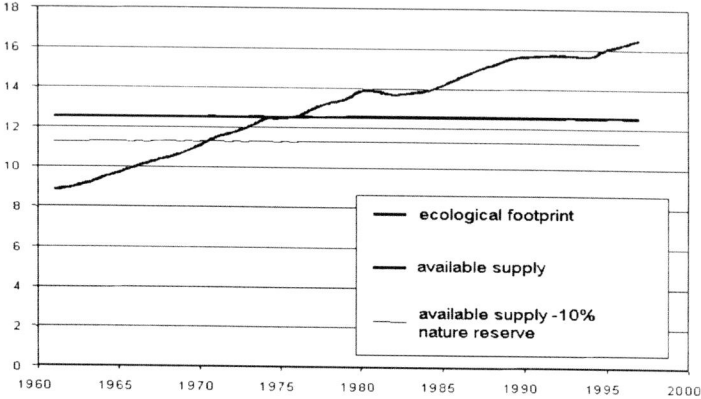

Figure 1. World ecological footprint, 1960 – 1996 (source [10])

Based on the ecological footprint, people already need nearly 30% more resources than nature can offer. For comparison, in 1960 the ecological footprint of the world population was only 8,9 milliard ha, nearly 30% less than nature can offer. Figure 1 shows the total ecological footprint compared to the available supplies, from 1960 until 2000.

This example shows clear enough that there is urgent need to bring this numbers back to equality, or at least to prevent the difference of becoming larger. There are three factors to work on:

- *[Productive area / world]: One approach is to enlarge the productive area of the world. More effective agricultural technologies and the exploration of still remote* areas are doing a bit. Even the exploration of areas for living outside the earth is subject to discussion and research.

- *[Number of inhabitants / world]:* Reducing this number is mainly a problem of ethics, culture and wealth. The aspects of the problem strongly differ between the parts of the world.
- *[Ecological footprint / person]:* The ecological footprint is very complex to calculate, and is just a rough estimation. It depends on many factors like:
  - Economical situation (e.g. wealth, standard of living)
  - Culture (e.g. consumption behavior)
  - Ecological infrastructure (e.g. alternative power production)
  - Logistic situation (e.g. transportation, areas of population concentration)
  - Product behavior (e.g. low power consumption, nontoxic materials)

The ecological footprint is probably the factor in the equation that has the biggest potential for reduction, at least within the next decades. The basic requirement for the reduction of the ecological footprint is the ecological awareness of the people. But in times and areas of wealth it is difficult to reduce the consumption behavior, especially because this is directly linked to the economical success of a country. So it must be tried to minimize the impacts to nature also without much changing in the people's behavior. This is where the idea of environmentally sound products is settled. They are minimizing the impacts without dispensing with them.

1.2 Environmentally sound products

A definition for an environmentally sound product is: *An environmentally sound product causes minimized impacts to the environment during its entire life.* Of course, products are seldom absolutely sound to the environment, but if the environmental impact is minimized they are at least sounder than others. The impacts are always caused by the processes through which the product passes during its life. The four main phases of product life are

- *the mining of the raw materials,*
- *the production,*
- *the use,*
- *the disposal or recycling.*

To determine the total impact of a product, all processes must be examined and the emissions must be calculated. To minimize the impact, this must be done before the product is actually manufactured. The characteristics of a product must be optimized during the product development.

This is a very complex task, mainly because of the high number of possible processes and the special knowledge that is important for each process to calculate the emissions.

Another problem occurs after the emissions are actually calculated: emissions can be of different types and they have different units. So they are not comparable and cannot be simply added together. This causes the need for a valuation of the emissions to make them comparable and addable.

These circumstances are directly resulting in a demand for special methods and tools for developing environmentally sound products.

2. BASIC REQUIREMENTS

Methods and tools for the development of environmentally sound products must be accepted by companies and product developers to be successful. To be accepted, they have to fulfill a lot of requirements. Some of the requirements are very close to the aim, like the functionality, which ensures that it is really possible to develop environmentally sound products. But there are also more fundamental and general requirements, like the adaptability to changing boundary conditions or the intuitive usability. Only the consideration of all these requirements can assure the success of such methods and tools. This chapter just lists the important requirements and describes some of their aspects.

Besides the requirements, which are fulfilled by the quality of the methods and tools, there is another aspect that decides over the success: the motivation for the industry to use the new methods and tools. Without such a motivation nobody would use the new methods and tools, and the quality and fulfillment of the requirements would be irrelevant. Chapter 3 is discussing the motivation for the industry to use the methods and tools, which is similar to the motivation of developing, producing and selling environmentally sound products.

2.1 Functionality

The most basic requirement is the functionality. The functionality is satisfying, if the methods and tools can do what they are meant for. In this case, it must be possible to develop environmentally sound products with the new methods and tools, and that in the desired accuracy. To define the functionality, also the question of "how" environmentally sound products should be developed must be answered. There are a lot of different approaches. All of them are based on an ecological assessment, but they can vary immensely in the level of detail.

Some approaches are handling the problem on a very low level of detail. An example is the estimation of ecological impacts of a product just by the material and mass of a product. In certain cases, the accuracy can be sufficient. Approaches with a low level of detail are most practicable in the early stages of product development.

Other approaches try to consider every emission and every eventuality in the behavior of a product. An example for a very detailed assessment method is the lifecycle assessment (LCA). It is more applicable to the later stages of product development, because detailed product and process data must be available.

See [2; 7] for a description of different approaches for ecological assessment.

2.2 Adaptability

The adaptability is a requirement that ensures the functionality for different boundary conditions. For example, the development of a household appliance needs different considerations than the development of a car, and the consumption of power has a different priority in different parts of the world. Some examples for conditions that require adaptation are:

- *type of product,*
- *country,*
- *laws,*
- *different views on weighting impacts,*
- *location,*

- *consumer behavior,*
- *and others.*

The methods and tools must be able to react on changes of such boundary conditions, which usually results in changes of the functionality.

2.3 Interdependency

The development of products depends on much more influences than the environmental soundness, like:

- *functionality of the product,*
- *economical aspects,*
- *design, fashion,*
- *consumer demands,*
- *available resources,*
- *competitive products,*
- *and others.*

All these factors decide over the success of a product. The problem about it is that they most often work against each other, for example:

- *a more fashionable product costs more,*
- *a more functional product consumes more energy,*
- *and others.*

The optimum depends on the importance of the different factors, and it varies for different products and markets. Methods and tools considering purely the environmental aspects of a product cannot be sufficient for a complete product development. It must be possible to consider all factors simultaneously and adjust them to find the optimum. The interdependencies between the factors are causing the need for links between methods and tools for the development of environmentally sound products and methods and tools for other aspects.

All aspects of a product should be assessed, compared and optimized from a common database to avoid redundancies. The problem is that each kind of assessment (for example ecological assessment or structural assessment (finite elements)) requires different kind and structure of product data. A common CAD database does not provide sufficient data and functions for most kinds of assessment.

2.4 Integrability

This requirement ensures that a company can integrate the new methods and tools into their existing structure.

On the methodical side it must be possible to extend existing product development methods with the new ones. The product developer must be able to consider the new facts together with existing methods.

On the technical side, the new tools must be able to be integrated into existing systems. Most of the data needed for the ecological optimization is created somewhere during the product development, and the recreation must be avoided. The existent data must be made available for the ecological assessment. Software interfaces to existing systems must be available. Such existing systems could be:

- *CAD Systems,*
- *PDM Systems and*
- *ERP (Enterprise Resource Planning) Systems.*

2.5 Usability

The usability allows the product developer to use the tools intuitively and efficiently. The user interface should be integrated into existing systems to avoid new and unfamiliar user interfaces.

The use of new functions, the creation of additional data and the visualization of results must be as easy as possible, unnecessary complexity must be hidden from the product developer. It must be easy for the product developer to find ecological weaknesses in a product, and he must automatically get useful information for optimization. This information could be linked directly links to the affected geometry or processes and generate hints where to start with the improvement.

An approach to visualize results of an ecological assessment is to start in a very low level of detail, like a single number or colors in the CAD or process model. From there, the user should be able to go into more detail in the areas of interest, like shown in figure 2.

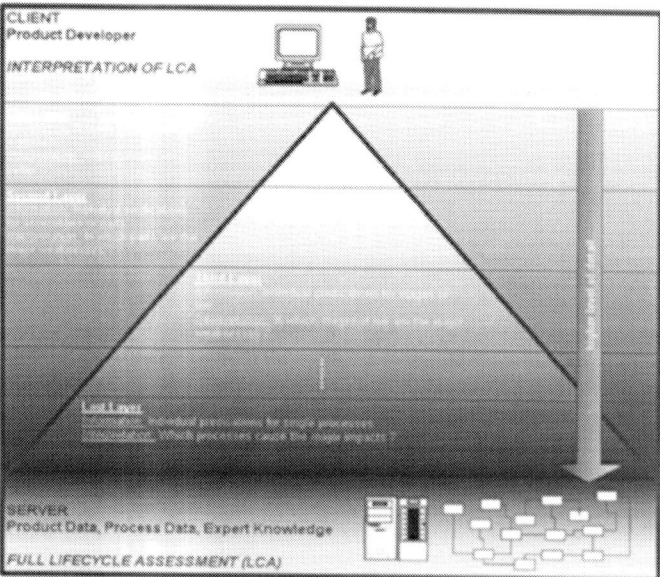

Figure 2. Levels of detail for user interface

3. MOTIVATION FOR INDUSTRY

Everything a company does must be driven by a motivation. The basic motivation of a company is to earn money. That can be on a short term by producing and selling products, or on a long term by research and the development of new products.

Unfortunately, the development of environmentally sound products most often stays in conflict with the economical aspects. The experience shows that products that are environmentally sound are more expensive to develop and produce, and therefore more expensive for the consumer. This effect must be compensated by a positive motivation for such products. The motivation could come from a higher sale volume by:

- *a good reputation of the company for developing environmentally sound products,*
- *the environmental awareness of the customers,*
- *awards for good environmentally behavior of products,*
- *economical advantage of green products due to technical aspects (e.g. power saving),*
- *economical advantage of green products due to political decisions (e.g. tax savings),*
- *"green" advertisements,*
- *and other aspects.*

The methods and tools can just help to make the development possible or easier, the motivation must be generated elsewhere.

There are lots of ways to increase such a motivation. One important factor is again the ecological awareness of the consumer. Ecological sound products can be made fashionable, for example by advertisements. This again influences the behavior of the consumer, and maybe even his ecological awareness.

One of the most powerful ways to increase as well the motivation for customers to buy ecological sound products as the motivation for companies to produce them are governmental decisions. A common way is the introduction of awards or tax savings for ecological sound products.

But such regulations are not without uncertainties about the ecological soundness, because they are usually reducing the product life and thereby increasing the amount of products produced. More material and energy for production is needed, and more waste produced. This can result in more damage to the environment, even if the new products consume, for example, less power.

Ingenious awards and tax savings for ecological soundness of products must consider the entire life cycle of a product (not only the use stage) in combination with the effects to the service life and amount of products produced.

4. THE SFB392 APPROACH

The methodical approach of the SFB392 is that the product development consists not only of the product design, but also of the life cycle design. The minimization of the impacts to the environment is done during the product development in an iterative approach: The impacts of a product or its parts can be analyzed for all stages in the life cycle, only on basis of virtual product- and process data. This analysis helps to find major weaknesses in the ecological behavior of the product. These weaknesses can be eliminated in the next iteration. Different versions and configurations of products can be analyzed simultaneously to decide for the best option. The assessment method used in the SFB392 is the most detailed one, the LCA (Life Cycle Assessment).

Under these aspects common Software used for product design, like 3D-CAD systems, is not sufficient to allow the development of environmentally sound products. Therefore a Product Development Environment (PDE) is being developed which considers the requirements and aspects discussed in chapter 2.

4.1 Overall architecture

The architecture of the PDE is based on a client – server concept. The client applications are used by the product developer to create the CAD data and the life cycle data, and to find the ecological weaknesses and react on them. The server applications are managing the data and providing the basic methods for the ecological assessment. They are also acting as the integration platform of the PDE, where client applications can be attached to a common interface. Figure 3 shows the client – server concept with the applications used in the PDE.

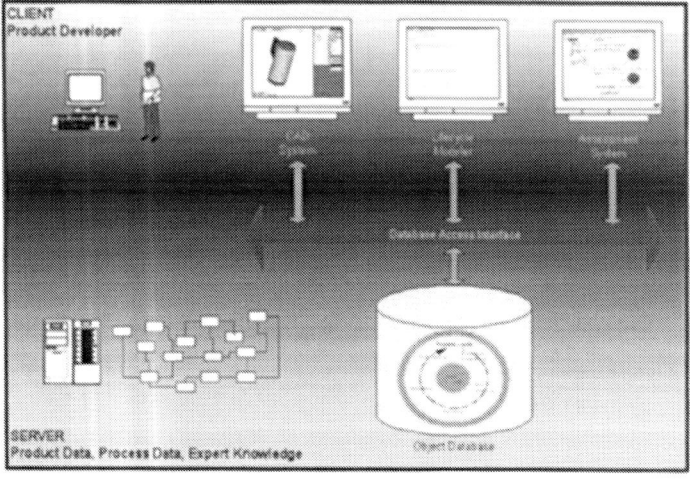

Figure 3. Architecture of the PDE

The following applications are currently used within the PDE:

- *Clients:*
  - *CAD System Pro/Engineer*
  - *Life Cycle modeler*
  - *Assessment System*
- *Servers:*
  - *Object Database System Object Store*
  - *Integration Interface CORBA*

A more detailed description of the architecture of the PDE is given in [1; 5].

4.2 Compilation of expert knowledge

The development of environmentally sound products requires much more information than a common product design. On the first view there are three groups of information that must be handled:

- *Product data: the geometry and structure of a product.*
- *Process data: the processes of the product life cycle.*
- *Process knowledge: how to do an ecological balance.*

Since all the information is linked together, it makes sense to handle all of this information in one system. An object oriented database system is used to manage the information. The object orientation has some advantages that are very appropriate for this purpose:

- *It helps to control the complexity of the information.*
- *It can implement methods that are used for the process knowledge.*

For the implementation in the database system, the information is divided into two groups, the information model and the instances.

4.2.1 The information model

The creation of the information model is driven by the process knowledge. It defines the kind and characteristics of processes, mainly what kind of data is needed and how the ecological impacts are determined. Since it defines which product- and process data is needed, it defines also the structure of product- and process data.

The information model contains mainly the following constructs:

- *Classes for product data.*
- *Classes for process data.*
- *Methods for ecological balance (part of the classes for process data).*
- *Interdependencies between the classes (associations, generalizations).*

To keep the information model flexible and open for enhancements, it is divided into different parts that are shown in figure 4.

The core model contains all the classes for the product data. It is mainly a reflection of the CAD model. Usually it needs no enhancement or adaptation.

The partial models are each responsible for a stage or group of processes in product life. They conclude the classes for process data, and thereby the methods and knowledge for the ecological balance.

The ecological balance itself is represented by a set of numbers, the inventory data. They are calculated by the methods of the partial model classes. They are a direct result from the data stored in the entire information model. The ecological assessment is based on the inventory data.

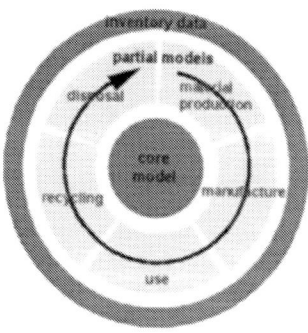

Figure 4. The information model

The information model is the basis for the functionality and the power of the PDE. It defines which processes of product life can be considered and how the ecological impacts are determined. It is also responsible for organizing and managing all the product- and process data that is created during the product development.

To adapt the PDE to certain environments (factories, types of products, etc.) the most important step is to adapt the information model. Every other part of the PDE has to satisfy the needs of the information model.

Since there are theoretically uncountable kinds of processes that a product can pass during its life, only an exemplary set of partial models is created in the SFB392 that covers typical processes for all stages of product life.

The information model is created by the SFB392 invented object oriented modeling language CoOM (*Cooperative Object Oriented Modeling*), which consists of notations for process modeling, static modeling and the verification of models [1; 3-6; 8-9]. The modeling language is presently changed to the international standard notation UML (*Unified Modeling Language*) [11]. A more detailed description of the information model of the SFB392 and information modeling in general can be found in [1; 3-6; 8-9; 11; 13].

4.2.2 The instances

While the information model is part of the PDE, the instances are a result from the work with the PDE. They are the product- and process data created during the product development, and they represent a certain product. Technically, this data are instances of the classes that are defined in the information model – they are objects. The objects are stored and managed by the object oriented database system. They are created from the CAD system or the Life Cycle Modeler. The assessment tool can use the inventory data that is calculated by these objects – the basis for the ecological assessment.

During the use of the PDE, only the instances are used. But the kind, structure and functionality of the instances are defined by the information model.

4.3 Product Development Methods

The development of environmentally sound product requires enhancements not only in the tools, but also in the methods. The product developer must integrate the application of the PDE into the development process. The application of the PDE is settled in the later phases of

D002/001 © IMechE 2002

product development, because it requires a high level of detail in the product and process data.

The use of the PDE is based on a very iterative method. In a first version of a product, the ecological weaknesses and their sources can be found with the assessment system. Then other variations of the product can be assessed by changing the appropriate product and process data, until the optimum is reached.

4.4 Adaptation of the functionality

The need to adapt the information model, and thereby the functionality, of the PDE to the special boundary conditions of products, companies, countries etc. was discussed previously. The client applications must supply this functionality to the product developer.

This problem cannot be solved with a static set of user interfaces and functions that are available to the user. Instead, the client applications must react on the changing structure of the information model.

For example, if the possibility to assess processes for surface hardening is added to the information model, the life cycle modeler must automatically supply a user interface for creating such a process, and the assessment system must automatically be able to include the new kind of processes into the ecological assessment. It would not be acceptable to change the code of each client application and recompile them each time the information model is changed. Possible approaches to avoid this are:

- *Each client application reads a central configuration file at startup and adjusts the user interface to the functionality described in the configuration file. The configuration file can be created automatically each time the information model is changed.*
- *Each client application scans the information model directly in the database system at startup and adjusts the user interface to the specified functionality.*
- *The functionality of the client applications can be adjusted with special tools or commands within the program.*

The interface to the database system must consist of a number of universal functions that are applicable to all possible versions and extensions to the information model.

5. SUMMARY, CONCLUSION

It is obvious that there is a need to preserve our environment. One approach is to minimize the impacts that are caused by our way of life. Since the products that we are using cause most of the impacts, a promising start is to optimize the ecological aspects of the products. This results in what is called environmentally sound products.

Next to the need and possibility to design environmentally sound products, a motivation for the industry to utilize this possibility is needed. This can mainly be reached by a demand on the market for environmentally sound products. Ecological awareness and governmental regulations are needed for that purpose.

How to really make products as environmentally sound as possible is the main topic of research within the SFB392. Tools and methods are developed that are helping to optimize the ecological behavior of a product during the product design. The tools and methods are based on a full Lifecycle Assessment (LCA).

The approach of the SFB392 considers all stages in product life, and it is theoretically possible to consider every process in the ecological assessment. The component based structure of the PDE and the information model allows a team of experts to implement their knowledge. It also allows the adaptation of the PDE to different boundary conditions, and the integration into existing system structures.

An evaluation done by experts in product development, psychologists, software specialists and about 30 test persons (mostly students) has proved the approach to be functional.

Despite these successes, there are still some problems to be solved and some requirements to be fulfilled. Most of the problems are a direct result of the high degree of detail the approach claims. Some fields that require work to do are methods and tools for the:

- *Creation and adaptation of information models*
- *Adaptation of the applications to the information model*
- *Definition of system boundaries*
- *Integration into existing system structures*
- *Integration into existing company structures*
- *Implementation of professional software systems*

These aspects will be considered in the next version of the PDE that is actually being developed. While aiming a complete and detailed ecological assessment, it has also a concept that allows a practical application in industry.

REFERENCES

1   Leibrecht, S., Pham Van, T.N., Anderl, R., 2002, "Integration of Expert Knowledge into the Development of Environmentally Sound Products "; *Proceedings of the Fourth International Symposium on Tools and Methods of Competitive Engineering – TMCE 2002*, Wuhan, China. (Pending)

2   Atik, A., Schulz, H., 1999, "Simplified Evaluation Methods for Development of Environmentally Sound Products in Early Design Phases"; *Proceedings of 6th International Seminar on Life Cycle Engineering*, Kingston, Ontario, Kanada.

3   Anderl, R., Daum, B., John, H., Pütter, C., 1998, "Information Modeling using Product Life Cycle Views"; *Proceedings of the 3th International Conference on the Design of Information Infrastructure Systems for Manufacturing – DIISM '96*, Fort Worth, Texas, USA.

4   Anderl, R., Daum, B., John, H., Pütter, C., 1998, "Integrated Development of Conceptual Models for product and Environmental Information Products"; *Proceedings of the Product Data Technology Days – 7th Symposium*, Watfort, Great Britain.

5   Anderl, R., Daum, B., John, H., Pütter, C., 1997, "Co-operative Product Data Modeling in Life Cycle Networks"; *Proceedings of the 4th CIRP International Seminar on Life Cycle Engineering*, Berlin.

6   Anderl, R., Daum, B., John, H., Pütter, C., 1997, "Information Modeling for Environmentally Sound Products"; *Proceedings of the International Symposium on Global Engineering Networking*, Antwerpen, Belgium.

7   Atik, A., 2001, "Entscheidungsunterstützende Methoden für die Entwicklung umweltgerechter Produkte" , *Dissertation TU-Darmstadt*, Shaker-Verlag.

8   John, H., 2000, "Modellierungstechnik zur Integration von Prozesswissen in ein Produktmodell", *Dissertation TU-Darmstadt*, Shaker-Verlag.

9   Pütter, C., 2000, "Kooperative Informationsmodellentwicklung: Grundlegende methodische und informationstechnologische Aspekte" , *Dissertation TU-Darmstadt*, Shaker-Verlag.

10   Loh, J., (Editor), 2000, "Living Planet Report 2000", World Wide Fund For Nature (WWF), Gland, Switzerland.

11   Jacobson, I., Booch, G. Rumbough, J., 1999, "UML – Unified Modeling Language (Version 1.3)", Rational Software Corporation, Cupertino, California, USA.

12   ISO 14040, 1997, "Environmental Management-Life Cycle Assessment – Principles and Framework", CEN European Committee for Standardization, Brussels.

13   ISO 10303-11, 1994, "Industrial automation systems and integration: product data representation and exchange. Part 11: Description methods: The EXPRESS language reference manual".

A web-based tool for design for sustainability of made-to-order products

P NORMAN
Department of Chemical and Process Engineering, University of Newcastle-upon-Tyne, UK

Abstract

Eco-design has been embraced by mass production industries such as packaging and automotive. These industries are relatively simple to evaluate in comparison with made-to-order (MTO) product systems such as offshore oil and gas platforms. During MTO design, engineers produce design specifications and select equipment subject to cost, reliability, weight, safety and spatial envelope. Recently, environmental performance and design for sustainability has been added to this list of criteria. To assist engineers, a Web-based tool is under development to help determine the environmental costs and benefits associated with these design decisions and to aim for sustainability. The work takes a 'systems' view of projects to design and build MTO products and involves identification and development of environmental performance criteria for design, using suitable examples as 'test beds'. The tool addresses the total life-cycle involving: plant materials extraction and processing, fabrication, operations, out-of-envelope conditions and decommissioning.

1 INTRODUCTION

This paper describes a web-based tool to support Design for Sustainability (DfS) within the Made-to-Order (MTO) industries; ie. those organisations involved in the construction and operation of complex, essentially one-off or short-run engineered systems such as power and processing plant, offshore platforms, ships and military aircraft. It is aimed at design engineers who may need to consider life-cycle environmental performance during conceptual and detailed design. The tool provides support for designers who are not necessarily expert in environmental and sustainability issues and provides guidance for decision making in the form of Environmental Performance Criteria which is supported by a knowledge base. The tool is organised into product life-cycle stages and a facility for storing and organising data generated by the user is proposed. The tool is intended to support the demonstration of Best Practice and contribute data to the Environmental Impact Assessment (EIA) process. It also presents a method by which contractors can demonstrate the fulfilment of environmental requirements handed down from clients who implement ISO 14001.

The requirement for such a tool follows on from a comprehensive survey of designers and managers in two major UK engineering companies (1). Results from the survey indicated that designers were generally aware of their responsibilities in terms of sustainability but they felt that the guidance and tools to help them were either too specialised and complex or were non-existent. There was significant support for the suggestion that better tools, especially those aimed at the non-expert, would enable sustainability to be integrated into the mainstream of the design process to be considered in parallel with the more traditional design concerns such as functionality, economic viability, safety and so on.

Barrow (2) noted that expert systems offered great potential for Environmental Impact Assessment (EIA) and environmental management. Antunes and Camara (3) proposed that for a computeraised EIA system to be useful and comprehensive it should be built using s decision support philosophy with an expert component. Their system – Hyper AIA – has the capability to incorporate quantitative and qualitative data and perform mathematical as well as logical operations in order to assist the decision maker in the selection of desirable alternatives. Navigation in this system is performed through a series of menus relating to the three main EIA tasks: identification, prediction and evaluation. Geraghty (4) reviewed a number of expert systems that have been developed as environmental assessment tools. Most were prototypes but their utility was noted. Mercer (5) proposed that there was a case for a computer-based system of general applicability that could function as a primary assessment tool in order to perform a quick assessment of a proposal. He suggested the tool should be used at the conceptual design stage to propose alternatives with less environmental impact. Mercer recognised that the nature of assessments was a qualitative one, relying on heuristic problem solving and proposed an expert system approach. However, it should be noted that none of these systems are aimed at providing whole life-cycle support for Design for Sustainability aimed at bringing the process into the mainstream of product design.

In specifying the requirements for the DfS tool proposed in this paper it was decided that it should not take the form of an 'expert system', as this would make the tool too generic and complex. Rather, it should be a structured information and guidance resource backed up with access to appropriate functional analysis tools and a mechanism to record design decisions, results, actions and comments. This would then allow designers the freedom to exercise their judgement and apply trade-offs. Implementation in a web-enabled form allows the tool to be modified or added to with the minimum of formality. The tool is organised loosely around a systems engineering approach and the main elements of systems engineering are referred to where appropriate.

2 SYSTEMS ENGINEERING AND THE DESIGN PROCESS

A system is comprised of sub-systems and system elements (5). It will operate with an external environment if it is an open system whereas a closed system can operate independently. The environmental surroundings of the system may exist in local, regional and global terms. For example, water discharge from an offshore platform can have a local effect whereas gas flaring results in carbon dioxide production which has a global impact.

System, sub-systems and system elements are often organised as a hierarchy. Interactions exist between:
- the system and the environment
- the system and interrelated systems
- relevant (and irrelevant) processes

Every system has inputs of material, energy and information. Outputs may be a mixture of useful and waste streams and are also categorised as material, energy and information.

Manufacturing Focus

Object-oriented modelling of deep drawn tailored blanks

J ULLRICH and **P GROCHE**
Institute for Production Engineering and Forming Machines (PTU), University of Technology-Darmstadt, Germany

Abstract

The Collaborate Research Centre (CRC) 392 "Design for Environment" at Darmstadt University of Technology supports designers with suitable tools and methods which allow them to develop environmentally sound products within technological and economic constraints.
This paper focuses on the computer-aided modelling of deep drawn tailored blanks. Therein, object-oriented programming is used to analyse the process sequence with respect to environmental effects. In addition to previous work by the CRC 392, experimental results are used to refine the computational model. A new item is the creation of a composite partial model (CPM), where models of different processes are interconnected. The main objective of this work is to proof the applicability of elementary formulas of forming processes in the CPM. An outlook is given to clarify the limitations of the theory and propose possible enhancements in the future.

1 INTRODUCTION

Over the last decades there has been a substantial increase in the environmental awareness of both manufacturer and consumer. Accordingly, legislation saddles the designers with environmental obligations the products have to fulfil [1]. Life cycle assessment (LCA) is a state-of-the-art method to compare consumptions of raw materials and energy of products.
The aims of LCAs comparisons are to inform the consumer and to reduce energy consumption. Recently, LCAs become increasingly relevant for managerial optimisation of products, processes, or even entire production facilities [2].
A major problem for the first LCAs was the absence of universal standardisation that defines the system boundaries. The absence of standard criteria in preliminary studies led to the

comparison of incompatible data [3]. To counteract this development, the ISO 14000 was initiated in the mid-1990s as an orientation guide for creating LCAs. The implementation of a LCA does, however, pose some critical problems. It requires, for instance, detailed knowledge about all life cycle stages of a product. An exact evaluation is often hindered by the fact that the production process has not been stringently defined. Another problem is, that an ISO 14000 consistent LCA is sometimes not economically convenient [4]. Furthermore, LCAs usually restrict to products which are already on the market. The most significant benefit for the environment therefore consists in the realisation of a prospective LCA. However, easily applicable methods, which regard holistic and prospective products and processes with respect to environmental criteria, are completely absent.

Current research at the Collaborative Research Centre (CRC) 392 focuses on the design for environment, methods, tools and instruments at Darmstadt University of Technology. The CRC 392 aims to provide designers with suitable software tools which allow them to design environmentally sound products within technological and economic constraints. A snapshot of the current research software is given in figure 1.

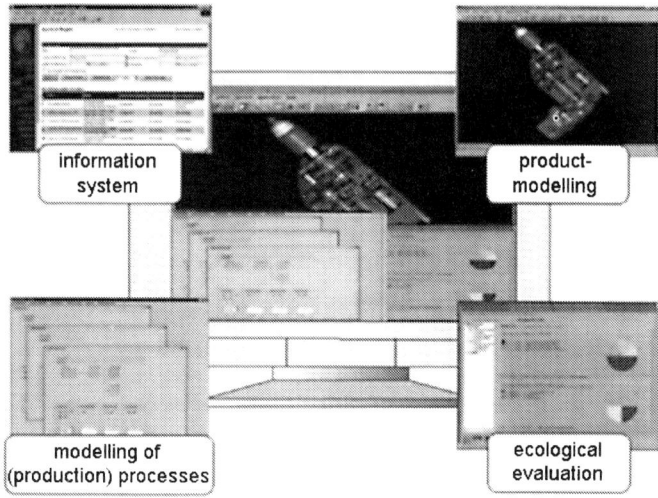

Figure 1: Snapshot of the software tool, developed by the CRC 392 [5]

1.1 The Activities of the Collaborative Research Centre (CRC) 392

It is necessary to consider all stages of the life cycle for the development of environmentally sound products. Environmental pollution is generally the result of production processes, use, recycling, and/or disposal of a product and its components. These processes and their energy consumption and material flux need to be analysed and quantified in order to undertake a holistic evaluation in terms of a product-related LCA. This procedure targets the analysis of existing products and requires computer-aided LCA-techniques. The integration of LCA methods into the product development requires detailed knowledge about all stages (such as production, use, etc.) and their interrelations as well as environmental effects.

The production processes examined in this publication are the basic techniques of production engineering as given by DIN 8580 as well as techniques for the production of plastics. A generalized picture of a life cycle is presented in figure 2.

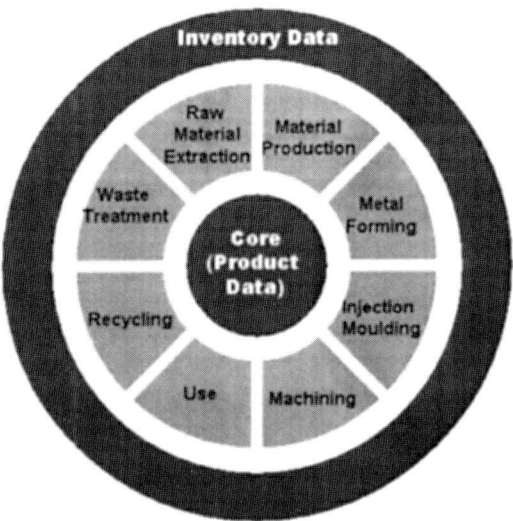

Figure 2: Information model of the CRC 392 [5]

So far, the CRC 392 has primarily analysed household appliance with a further focus on consumer goods. From an ecological perspective of view, the automobile is a particularly object of interest due to its high amount of primary energy over its whole life cycle and the potential to economise a lot of energy. As far as the whole life cycle of the product is concerned, 80 to 90% of the entire energy-need is consumed during product use in the form of fuel [6]. A feasible method to decrease fuel consumption is given by the reduction of car weight. This is achieved by choosing alternative materials as well as by innovative design principles, combined with new manufacturing technologies, or by combination of production processes (see next section). The production of deep drawn tailored blanks in the automotive industry requires this kind of a combination of production processes which has been introduced in the mid-1980s. For the CRC 392 it is of particular interest to what extent the use of deep drawn tailored blanks (over the whole life cycle) is ecologically and economically lucrative. Consequently, there is a strong need to develop a partial model for deep drawn tailored blanks.

In the beginning, only a rough vision of the product structure is needed for the modelling of the production processes and product shape. First, materials, the type of the semi-finished product, and dimensions are selected. Then, all applicable production processes, as well as sub-process chains, need to be determined. (A combination of forming and shearing production of a component might be, for instance, an alternative to a metal cutting production.)

1.2 Tailored Blanks

Welded blanks comprising different materials, thickness or coatings are known as tailored blanks. Tailored blanks provide the designer with new possibilities for designing components. Besides allowing for reduction of weight, tailored blanks also allow possible a stress-compatible construction of components by way of choosing specific combined sheets. In doing so, stiffness and strength of the vehicle-structure can be improved significantly. Moreover, fewer components simplify the product logistics while the accuracy in size and shape of the parts or assemblies is increased at the same time. The elimination of stiffeners, and thereby of overlapping joints, improves the corrosion resistance of the assemblies and eliminates the need for sealing work and sealing compounds. Not only does this cut costs, but it is also advantageous for subsequent recycling. Furthermore, it is worth mentioning that tailored blanks can be formed in the same way as normal blanks [7].

The processes that are taken during deep drawing of tailored blanks are a combination of the production processes of welding, deep drawing, and shearing. Within the last three years, all these processes have been analysed and modelled in an object-oriented way [8, 9]. An innovative aspect of this work is the combination of the different processes to a composite partial model (CPM) and more important, the modification of the chronological order of the processes. Conventional deep drawing processes start with deep drawing of sheets, followed by shearing and welding. In contrast to this procedure, deep drawn tailored blanks are first welded before deep drawing and shearing is done.

Generally, tailored blanks can be formed in roughly the same manner as normal blanks. It should be noted, however, that the welded joint must be arranged in such a way as to impede it from migrating in the transverse direction. Beside, there is a risk for the welded joint to become superelevated because of the different thicknesses of the blanks.

2 RELEVANT PROCESSES FOR DEEP DRAWN TAILORED BLANKS

2.1 Welding of Single Sheet

According to DIN 8580, welding is defined as the merging of components consisting of similar or unequal base material. During the welding process, either the temperature in the welding area is raised or the pressure on the components is increased to allow local melting of the components. Occasionally, additives have to be added [10]. The welding procedures which are commonly used for the joining of tailored blanks are laser and mash welding. Laser welding allows the assembling of very thin materials, because the thickness of the welded joint is only a few tenths of a millimetre. Laser welding can be done without additives. A clear advance of this technique is, that even heat sensitive materials can be joined due to the restricted heat exposure of the components during the welding. Accordingly, the heat exposure during mash-welding is even lower, but this technique lacks from the sheet overlapping commonly found after welding.

The welding process has already been modelled in work done by the CRC 392 [11]. The required energy for welding is calculated according to the following formula:

$$E_{weld} = \frac{U_{weld} \cdot I_{weld}}{\eta} \cdot \frac{l_{weld}}{v_{weld}} + p_{0v} \cdot \frac{l_{weld}}{v_{weld}} \cdot 1,1 \quad [\frac{kg \cdot m^2}{s^2}] \tag{1}$$

where:

U_{weld}	voltage [V]
I_{weld}	strength of current [A]
η	efficiency of the welding source [-]
l_{weld}	length of the seam [m]
v_{weld}	welding speed [m/s]
p_{0v}	idling capacity [w].

In addition to energy consumption, possible sources of gaseous and particle pollution are also considered in the model. These emissions vary significantly depending on different welding processes.

2.2 Deep Drawing

In accordance to DIN 8582, deep drawing is a compression-tension forming process. It concerns a production process well suited for mass production, since high quantities can be produced in short intervals with few tools [12]. Deep drawing with rigid tooling is a large-scale application. It involves the use of drawing punches, a blank holder, and a die. In this process the blank is reformed, pulling it over the draw punch into the die. The blank holder prevents the blank from being wrinkled in the flange. The thickness of the sheet metal remains unchanged. The material usage of deep drawing varies between 50 and 75% and depends on the geometry of the blank as well as on the arrangement of the parts on the blank and the dimensions of the coil. Liquid or paste-like lubricant is ordinarily used to reduce friction and wear between blank and tool. The quantity of required lubricant depends on the geometric size of the blank's surface. Variations in lubricant viscosity are subject to the complexity of the forming operation. For instance, it is sometimes possible to reduce the number of steps in complex forming operations by using a highly viscous lubricant. The disadvantage of such a procedure is that a highly viscous lubricant requires significantly more cleaning agent for degreasing the surface of the sheet. From an ecological perspective, deep drawing processes cause the biggest environmental damage due to the use of cleaning agents.

According to eq. (2), the forming energy W_d can be determined from the maximum deep drawing force F_Z and the depth of draw h with sufficient accuracy [12].

$$W_d = 2/3 \cdot F_Z \cdot h \quad [Nm] \tag{2}$$

with h [mm] = depth of draw.

The formulas for the determination of the deep drawing force F_z according to the slab method will be discussed in detail later.

2.3 Shearing

According to 8580, "separation" compiles all processes designed to remove the cohesion of material locally. Among there, shearing has the biggest commercial relevance among all separating processes. In shearing, cross-sectional areas are typically classified into the notched zone, the shear zone, the fracture zone, and the burr zone. Cost-related factors (e.g. tool wear) are often most important in the shearing process, much unlike other sheet metal forming processes in which the quality of the material plays a major role [13]. Cutting oil is used to reduce tool wear. Using shearing in conjunction with other forming processes (e.g. deep drawing) makes additional "oil lubrication processes" unnecessary, since lubricants can serve the same purpose as cutting oil. Generally, shearing and forming processes are to be followed by cleaning and degreasing of the product.

According to ref. [12] the necessary shearing force is calculated with eq. (3):

$$F_s = l_s \cdot s \cdot 0,8 \cdot R_m \quad [N] \tag{3}$$

where:

l_s length of a cut [mm]

s   sheet metal thickness [mm]
R_m material tensile strength [N/mm^2].

The shearing work can be approximated by:

$$W_s = 2/3 \cdot F_s \cdot s \quad [\text{Nm}] \tag{4}$$

3 OBJECT-ORIENTED MODELLING

The model development of a product lifecycle is separated in two steps. First, the energy/ material flux is identified and modelled in accordance to the specified manufacturing process using the IDEF0 programming language. This also allows the hierarchical structuring of processes. Material flux is only taken into account if the mass is relevant in proportion to the amount of the total mass flow or if the mass flow poses a potential environmental threat due to its constituents (e.g. toxic substances).

Second, the resulting flow charts are mapped into object-oriented models with a certain display format suitable for evaluation by our electronic data processing environment. Object-oriented modelling allows to fraction complex systems into small units (black boxes). Consequently, a broad user base is enabled to master this technique, even without sufficient programming knowledge. Given that the functions are directly linked to the data, increments and modifications of the process flow are easy to manage. In advance to the final system set-up, the boundaries need to be fixed as to include parameters of all major process chains. In accordance to common conventions of life cycle assessments, the factors manpower and processes caused by the presence of the jobholder, such as commuting to and from work, can be disregarded [14].

The resulting process models are generally tailored to allow the analysis of technological, ecological and economical data. This data provides the user with knockout criteria for specified production processes, manufacturing costs of specific products, and inventory data.

Some important parameters regarding the processes of deep drawing and shearing are sheet thickness and sheet surface, and, depending on the process-type, the volume of the parts and the cutting length, respectively. Inventory data, which is determined by deep drawing and shearing processes, are energy consumption, consumption of lubricant or cutting oil and cleaning agent. Energy consumption results from the forming force, (specifically, the cutting force) and the degree of efficiency of the forming machines. The consumption of lubricants and cleaning agents is modelled by sub-classes with adjustable factors. The model parameters for welding include the cutting length, cutting speed, degree of efficiency, and power. Emissions which result from the process are gaseous and particle-shaped pollutants as well as electric energy.

4 APPROACH

4.1 Slab method

The basic theory for the partial models is provided by the elementary plasticity theory (slab method) [15]. Compared to other solutions, the slab method has the advantage of being easy to handle in terms and theory. This method approximates the kinematical movements of "slabs" in an analytic way. It was shown that the slab method often yields admissible results

[16]. Particularly integral aspects such as forces, moments, energy, and power often agree well with experimental results. In addition, the slab method provides formulas which permit to rank the impact of individual parameters. The sole input data for this method are the material tensile strength of the forming material, the coefficient of friction, and the geometry. The slab method is only applicable to pure materials. In all other cases, it is necessary to derive average values of the materials in advance.

The primary result of the slab method is the total force of the forming process. Therefore it seems reasonable to make use of this method in the object-oriented modelling, since the total force is sufficient to calculate the energy consumption of a forming process. However, it should be noted that a detailed analysis of the forming process requires the use of a more elaborate theory (e.g. finite element methods).

According to the slab method, the maximum drawing force F_z can be calculated from eq. (5) [15]:

$$F_z = (F_{id} + F_{RN}) \cdot e^{\mu \cdot \alpha} + \frac{k_{fmII} \cdot \pi \cdot d_m \cdot s_0^2}{2 \cdot r_R + s_0} \quad [N] \tag{5}$$

where F_{id} and F_{RN} are given by

$$F_{id} = 2 \cdot \pi \cdot r_{p2} \cdot s_0 \cdot 1{,}1 \cdot k_{fmI} \cdot \ln(R/r_{p2}) \tag{6}$$

$$F_{RN} = 4 \cdot \pi \cdot r_{p2} \cdot \mu \cdot p_N \cdot (R - r_{p2}) \tag{7}$$

where:

μ coefficient of friction at the drawing die radius [-]

α angle at the drawing die radius with $\alpha = \mu/2$ [-]

r_R drawing die radius [mm]

d_m median diameter of the cup with $d_m = d_1 + s_0$

d_1 internal diameter of the cup [mm]

s_0 sheet metal thickness [mm]

d_0 diameter of the blank at the beginning of the forming process [mm]

p_N blank holder pressure: $p_N = x_{pN} \cdot 10^{-3} \cdot [(\beta - 1)^3 + 0.5 \cdot 10^{-2} \cdot d_0/s_0] \cdot R_m$ [kg/m·s]

x_{pN} coefficient of the blank holder pressure (between 2 and 3) [-]

R_m material tensile strength [N/mm²]

r_{p2} radius of the flange in point 2: $r_{p2} = d_1/2 + s_0 + r_R$ [m]

$k_{fmI,II}$ median yield stress in point 1 and 2 respectively, depending on the material tensile strength [N/mm²]

4.2 Results

4.2.1 Deep drawing experiments

Several experimental set-ups were built at the Institute for Production Engineering and Forming Machines (PTU) to study the deep drawing of tailored blanks and collect data for the computer-assisted follow-up studies. The main goal experiments determine the force displacement behaviour of both materials. During the experiments the behaviours of different sheet qualities were studied. The basic geometry was given by a circular blank with a 200 mm diameter and 1 mm thickness. ZstE340 (hard deep-drawing steel) and FeP06 (soft deep-drawing steel) were chosen as materials for the tailored blanks. Deep-drawing experiments were taken with a tailored blank and a pure blank as a reference. Each material comprised 50% of the tailored blank being welded by a laser weld of 1.5 mm width. The diameter of the punch was 100 mm. The experimental set-up is depicted in figure 3.

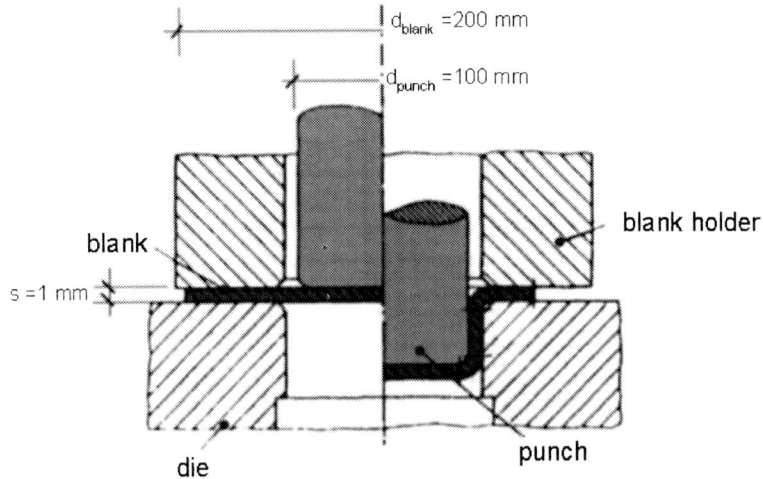

Figure 3: Experimental set-up for deep drawing

During the deep drawing process it was found that the FeP06 reaches a maximum force of 129,3 kN, while the ZstE340 reaches a maximum force of 151,7 kN. The maximum force measured on the tailored blank was 138,3 kN.

The time-resolved force curve was similar for all experiments, while the force maximum was reached at a penetration depth of 35 mm (see next section).

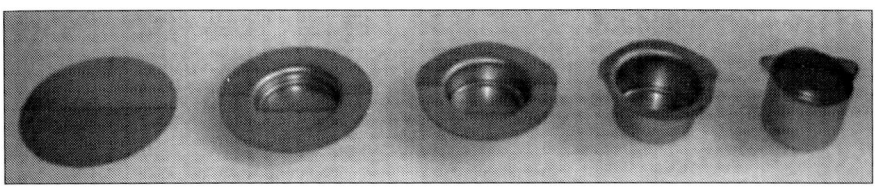

Figure 4: Stages of deep drawn tailored blanks

Even at slight draw depths, the soft sheet of the tailored blank forms visible lappets in the flange. Near the bottom of the punch the seam warps towards the hard sheet (the harder material trails the soft material). In contrast, the seam at the side panel runs in a slight s-curve towards the other direction (the hard material displaces the soft material). This effect is shown in fig. 4 and 5. A comparison of the forces reveals that the maximum force of the tailored blank does not lie halfway between the maximum force of both steels, but that the ZstE340 contributes 40% and the FeP06 contributes 60% of the maximum force of the tailored blank respectively. This effect might be well explained by the fact that the hard material displaces the soft material and consequently, the hard material requires less force for being formed.

D002/003 © IMechE 2002

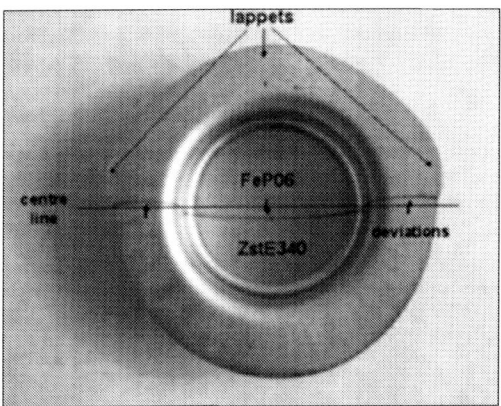

Figure 5: Tailored blank after the deep drawing experiment. The geometrical measures are as follows: diameter: 200 mm, sheet thickness: 1 mm, seam: 200 mm, depth of draw: 100 mm, weight: 247 g.

4.2.2 Calculations

Since it was not possible to compute the deep-drawing of tailored blanks according to the slab method (cf. chapter 4.1), the tensile strength of the hard and soft material was inserted separately into the formulas. In contrary to the experiments, the calculation results a single value for the maximum force of the punch rather than yielding a force-displacement-curve (see fig. 6). The tensile strength of the ZstE340 was set to 435 N/mm² and the tensile strength of the FeP06 was set to 304 N/mm².

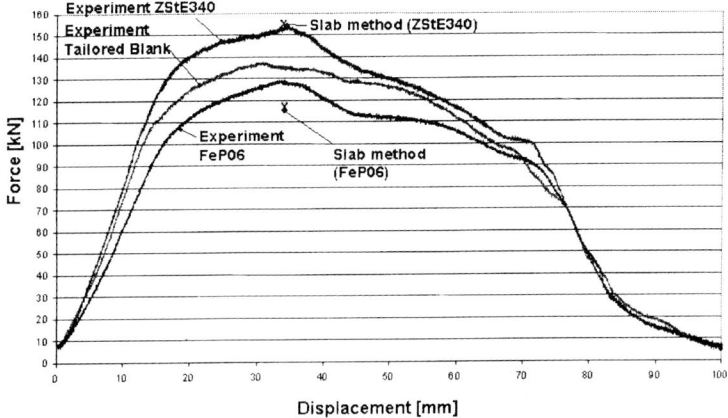

Figure 6: Force-displacement curve resulting from the deep drawing measurements described in section 4.2.1 in comparison to calculation results

According to the slab method, the FeP06 was found to reach a maximum force of 116.4 kN, 9,0 % lower than the value yielded by the experiment. The ZstE340 reached a maximum force of 151,9 kN and deviated by less than 1% from the experimental value.

The maximum force for tailored blanks might be well approximated by the mean value of the force maximum of both single sheets (eq. (8)).

$$F_{TB} = 1/2 \cdot \left(F_{ZStE340} + F_{FeP06} \right) \qquad (8)$$

From the results given in table 1, it can be concluded that the eq. 5-8 are well suited to approximate the force of the deep drawing of tailored blanks within computer-aided studies. This is the primary result of this study.

Table 1: Comparison of calculations with experimental results

Material	Tensile Strength R_m [N/mm²]	max. Force [kN]		Deviations [%]
		Experiments	Slab method	
ZstE340	435	151,7	151,9	0,2
FeP06	304	129,3	116,4	9,0
Tailored Blank	-	138,3	134,2	3,0

4.2.3 Energetic Aspects

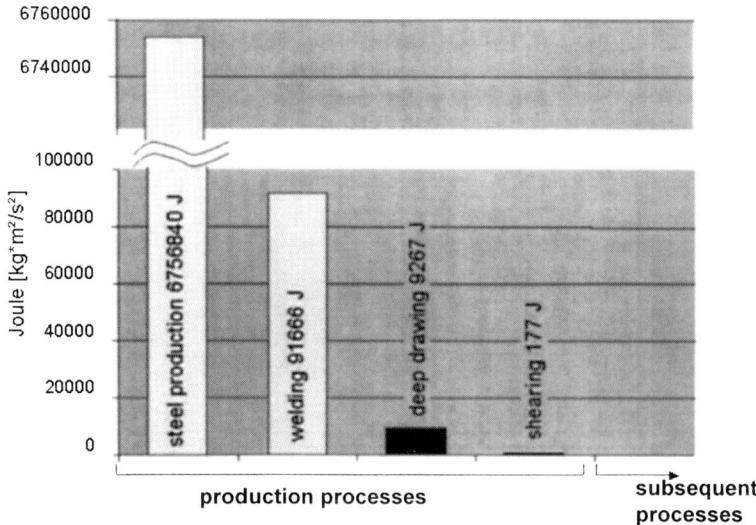

Figure 7: Energy consumption of tailored blanks according to the described geometry (diameter: 200 mm, sheet thickness: 1 mm, seam: 200 mm, depth of draw: 100 mm, weight: 247 g)

Given the basic equations (eq. 1-8), the energy consumption of metal production and forming processes can be determined with sufficient accuracy. Based on the results of the previous section, a new module for the deep drawing process of tailored blanks has been developed which is ready for the integration within the modelling software environment.

Calculations show that steel production is indeed the most energy-consuming step in the production chain of blanks. This is also the case for the production of deep drawn tailored blanks. Figure 7 illustrates the energy consumption for the production of a deep drawn tailored blank with the geometry shown in figure 3. The amount of energy for the material production was calculated according to a mass-dependent factor that accounts for the different phases of material production [17]. The subsequent process energy derived from the equations given in chapter 2. What immediately shows up, is the remarkable high amount of energy spent for welding. In contrast, the deep drawing proves to be 10 times more efficient, while the shearing energy is almost negligible.

5   CONCLUSION AND PERSPECTIVE

In this work, a new module has been developed for the computer-assisted modelling of the deep drawing process of tailored blanks. It was designed as an integrative part of the current modelling software environment of the Collaborate Research Centre (CRC) 392 "Design for Environment" at the Darmstadt University of Technology. This module is aimed to support product designers with data regarding the environmental effects of the production chain. Therein, the main aspects are the process energies, material consumption, degree of efficiency and the consumption and toxicology of additives.

An experimental set-up was built to measure time-dependent force curves of deep drawing of blanks and tailored blanks. These forces were compared with the results of the well-known approximate slab method within an error range of 0,2-9,0 %. Consequently, this method is suitable for predicting the energy consumption on the basis of forming forces. However, the test series is not yet completed and there remains a need to examine other parameters of tailored blanks like different sheet thickness and coatings. Emphasis should be placed on the construction of a special tool for use in the variation of different sheet thickness. Finite-element-simulations are planned to complement the experiments, thereby allowing for a significant reduction of experimental effort. Case studies of this kind are underway and will be presented in a forthcoming publication.

REFERENCES

1 **Klöckner, S.** and **Müller, D.** Umweltgerechte Produktentwicklung – gestern, heute und morgen. In *Umweltgerechte Produktentwicklung – Ein Leitfaden für Entwicklung und Konstruktion*, Deutsches Institut für Normung e.V. Beuth Verlag GmbH, Berlin Wien Zürich, 2000

2 **Schmidt, M.** and **Schorb, A.** Stoffstromanalysen in Ökobilanzen und Ökoaudits, Springer-Verlag Berlin Heidelberg New York, 1995

3 **Grüner, C.** and **Birkhofer, H.** Entwicklung umweltgerechter Produkte – Methoden und Organisation. In *Verantwortung Umwelt – Herausforderung für die Produktentwicklung der Zukunft*, Technical University of Darmstadt, Germany December, 5th. 2000, Darmstadt, pp. 1-8

4 **Schlotheim, G. Frh.v., Groche, P., Schmoeckel, D.** and **Wansel, A.** Object-Oriented Modelling of Select Metal Forming Processes, In Proceedings of HelSIE: Helsinki, Symposium on Industrial Ecology and Material Flows, Helsinki, Finland 2000, pp. 262-273. University of Jyväskylä, Jyväskylä 2000. ISBN 951-39-0783-X

5 **Anderl, R. Daum, B.** and **John, H.** Life Cycle Management: Voraussetzung für die Entwicklung umweltgerechter Produkte. In *Verantwortung Umwelt – Herausforderung für die Produktentwicklung der Zukunft,* Technical University of Darmstadt, Germany December, 5th. 2000, Darmstadt, pp. 33-39

6 **Pollmann, W.** Die Entwicklung umweltgerechter Produkte als Kernaufgabe in der Automobilindustrie. In *Verantwortung Umwelt – Herausforderung für die Produktentwicklung der Zukunft,* Technical University of Darmstadt, Germany December, 5th. 2000, Darmstadt, pp. 10-19

7 **Thyssen Fügetechnik** Tailored Blanks – Optimized Components in Steel Sheet, Information brochure, May 2001, Duisburg, Germany

8 **Schlotheim, G. Frh.v.** Objektorientierte Prozessmodelle zur prospektiven Sachbilanzierung umformtechnisch hergestellter Bauteile, Dissertation, Technical University Darmstadt, Shaker Verlag Aachen 2002, ISBN: 3-8265-9824-5

9 **Schiefer, E.** Ökologische Bilanzierung von Bauteilen für die Entwicklung umweltgerechter Produkte am Beispiel spanender Fertigungsverfahren, Dissertation, Technical University Darmstadt, Shaker Verlag Aachen 2001, ISBN: 3-8265-8724-3

10 **Ruge, J.** Handbuch der Schweißtechnik, Band II, Verfahren und Fertigung, Springer-Verlag Berlin Heidelberg New York, 1980

11 **Schiefer, E.** Technische, wirtschaftliche und ökologische Beurteilung von trennenden, fügenden und beschichtenden Verfahren, Technical Report, Darmstadt1999

12 **Schuler AG** Handbuch der Umformtechnik, Springer-Verlag Berlin Heidelberg, 1996

13 **Schmidt, M., Schorb, A.** Stoffstromanalysen in Ökobilanzen und Öko-Audits, Springer-Verlag Berlin Heidelberg, 1995

14 **John, H.** Modellierungstechnik zur Integration von Prozesswissen in ein Produktmodell, Dissertation, Technical University Darmstadt, Shaker Verlag Aachen 2001, ISBN: 3-8265-8377-9

15 **Groche, P.** Umformtechnik I, lecture notes, Institute for Production Engineering and Forming Machines, Technical University Darmstadt, Germany, 2001

16 **Kopp, R.** and **Wiegels, H.** Einführung in die Umformtechnik, 1. Auflage, Verlag der Augustinus Buchhandlung, Aachen, 1998

D002/003 © IMechE 2002

17 Wolf, B. Beurteilung des Recyclings bei der Entwicklung umweltgerechter Produkte, Dissertation, Technical University Darmstadt, VDI-Verlag Düsseldorf, 2001, ISBN: 3-18-333-801-7

Applications for eco-efficient surface machining with dry ice blasting

E UHLMANN, F ELBING, and **A EL MERNISSI**
Institute for Machine Tools and Factory Management, Technical University of Berlin, Germany

Abstract

New approaches in European industry have been brought about by a sensitisation of the public for environmental protection, by rising disposal costs and a new legal framework. Cleaning processes that are applied in almost all branches of industry are a potential health hazard for humans and the environment. The Institute for Machine Tools and Factory Management at the Technical University Berlin develops future-orientated eco-efficient cleaning technologies for hard surfaces and co-operates with companies to integrate these technologies into the process chains of production, maintenance and recycling. Other fields which one uses dry ice blasting are surface preparing and cutting. The paper presents results of system developments, customised process optimisations and fundamental research of dry ice blasting.

1 INTRODUCTION

The industrial surface treatment technology for hard surfaces is an important production step in manufacturing, maintenance, repair, and recycling processes. The European industry is forced to change the current cleaning processes not only due to technical aspects, but especially because of the new ecological and economical circumstances. The main reasons are increased environmental awareness, new environmental legislation, rising waste disposal costs, and increased sewage taxes.

At the moment, mainly chemical, mechanical and aqueous surface treatment processes are used. These processes are characterised by a lacking flexibility regarding contamination and basic material, an inadequate quality due to the changing demands, and by an abrasive and corrosive influence on the component to be treated. An energy and time intensive follow-up treatment, washing, as well as drying of the components is necessary and adds to the complexity and size of surface treatment equipment.

Moreover, chemical, solvent, and sound emissions lead to health risks for workers and environment. The contaminated sewage from surface treatment processes is often not treated and collected sufficiently. This leads to water and soil pollution in many places in Europe. There is an extensive loss of production caused by long down-times at disassembling, cleaning, and assembling the systems.

In 1997, for example, the production output rate of cleaning equipment of the German engineering industry was about Euro 0.9 billion, and the market share of new environment-friendly processes becomes more and more significant. Research is beginning to develop a number of new dry-cleaning processes, like for example laser, plasma, and environment-friendly blasting processes. The European industry has recognised the urgent need for action, and the cleaning technology is currently in a state of change. It can be assumed that the substitution of conventional cleaning technologies will continue in the next years and that the application of new technologies will increase steadily.

Dry ice blasting is one of these new technologies, although the industrial application of this technology has been tested since 1970 [1, 2, 3, 4]. Since the industry is lacking process data for most of the relevant applications, as well as data regarding possible changes of the material being processed, it cannot implement dry ice blasting, yet.

2 DRY ICE BLASTING

2.1 History
The basis for dry ice blasting was created 1930, when it succeeded for the first time to manufacturing solid carbon dioxide (CO_2) in the form of dry ice, i.e. in the solid state to manufacture [5]. After 1945 executed the U. S. Navy and Airforce the first blasting tests for cleaning with dry ice. To the commercial application of the procedure it came only after 1963. Patents from the years 1972 and 1977 describe still today the applied procedure version of blasting with dry ice pellets [6, 7]. The first industrial dry ice blasting systems were developed 1980 in the USA. Production and distribution of these systems took place in the start by the companies Alpheus and ColdJet. In the meantime existed world wide a multiplicity at companies, the equipment technology and services for dry ice blasting offer.

2.2 Process
Dry ice blasting is a pneumatic jet process that uses dry ice pellets as a one-way blast medium. Dry ice pellets consist of solid carbon dioxide (CO_2) at a temperature of -78.5 °C. The production of the blast medium is based on fluid carbon dioxide that can be won as a by-product from hydrogen, ammonia, and ethanol production, as well as from oil and gas refineries. Relieving the carbon dioxide to a pressure of 1 bar at a temperature of -80 °C generates dry ice snow. A hydraulic stamp presses the snow through a mould into a pelletizer. The pelletizer produces cylindrical dry ice pellets. Pellet parameters that effect the cleaning process are density, hardness, surface finish, shape, carbon dioxide content, and dimensions of the CO_2-pellets [8,9]. **Figure 1** shows the relevant parameters in dry ice blasting.

There are no residues of the solid blast medium left after the cleaning process, because the dry ice immediately sublimes when striking the surface, i.e. it attains the gaseous phase. This represents an advantage compared to other cleaning processes where the medium either requires complex processing, or involves costly disposal together with the impurities dissolved in the medium. The fact that the dry ice pellets turn to gas also means that no injection media remains are in the boreholes or the cavities of the machined parts. Subsequent cleaning or drying, as in other processes, is not necessary. Dry ice blasting does not produce any residue, is only slightly abrasive and corrosive, and flexible concerning different materials and contamination [10].

D002/019 © IMechE 2002

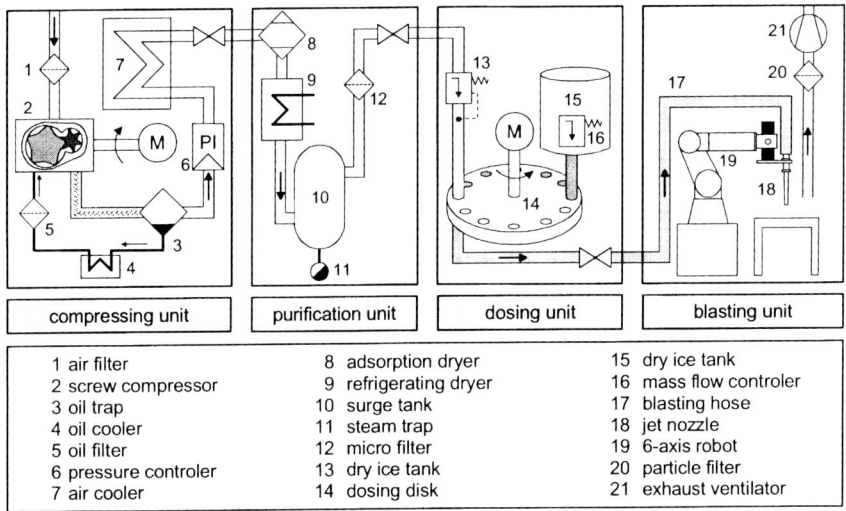

| compressing unit | purification unit | dosing unit | blasting unit |

1 air filter	8 adsorption dryer	15 dry ice tank
2 screw compressor	9 refrigerating dryer	16 mass flow controler
3 oil trap	10 surge tank	17 blasting hose
4 oil cooler	11 steam trap	18 jet nozzle
5 oil filter	12 micro filter	19 6-axis robot
6 pressure controler	13 dry ice tank	20 particle filter
7 air cooler	14 dosing disk	21 exhaust ventilator

Figure 1: Dry ice blasting parameters

During the technical process, the pellets are collected in a storage element and injected into an air stream by means of an impeller wheel used as a dosing device. The pellets are accelerated by the compressed air stream through a laval nozzle and depart from the nozzle at almost the velocity of sound [11].

Compared to other processes dry ice blasting can, in numerous ways, reduce costs. On the one hand, the reduced waste volume, compared to sand blasting or chemical processes, minimises disposal costs. On the other, work time is saved because application can take place on site without having to remove sensitive parts. Machining times are distinctly shorter than those when sand blasting is used. Time savings can be as high as 50 to 75 % when blasting molds [12]. This less aggressive method also serves to prolong the service life of the moulds. One disadvantage inherent in jet processes that are operating in fully automatic mode is that they expose operators to extreme working conditions. Sound pressure levels of up to 125 dB(A) were measured for dry ice blasting during machining with equipment which was not hermetically sealed. Therefore, the process should be conducted with hermetically sealed equipment and staff should be equipped with hearing protection. Moreover, the CO_2-concentration in the surrounding air may be elevated.

2.3 Active Mechanism

The active mechanism in dry ice blasting is based on a combination of thermal and kinetic energy input, as well as sublimation energy [13, 14]. The thermal energy supplied during the cleaning process leads, on the one hand, to a regional undercooling of the part where the pellets strike the surface. As a result, elasticity is lost and the adhering coating becomes embrittled and shrinks while forming cracks. Due to the different thermal expansion coefficients of the coating and the substrate, the bond with the substrate dissolves when the adhesive energy is exceeded. The coating partially chips off. On the other hand, the kinetic energy of dry ice pellets and air stream contributes to the removal of the coating [15, 16]. The sudden increase in volume resulting from the sublimation when the pellets strike the surface of the part supports the process **Figure 2**. Gas flows underneath the adhering coating. The removal of material is hence based on a combined thermo-mechanical effect.

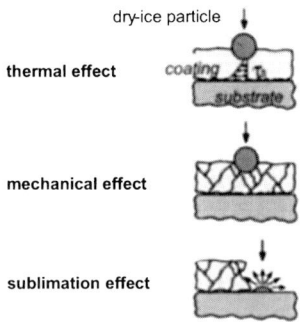

Figure 2: Active mechanisms of dry ice blasting

3 EXPERIMENTAL INVESTIGATIONS

3.1 Experimental Set-Up

Experimental investigations on dry ice blasting were conducted with a compressed air system by Kaeser Kompressoren, Coburg/Germany, which comprise a stationary single-stage screw compressor DSB 170, an equalising vessel with a volume of 0.4 m³, a steam trap Eco-Drain 13, and a micro filter element FX to collect oil and water. The screw compressor with a nominal motor performance of 90 kW provides for a maximum blast pressure of 13 bar and a maximum volume flow of 16 m³/min. The mobile blast system used, CryoMax II of Linde, Munich/Germany, moves the blast medium according to the pressure principle. For guiding the jet nozzle in the machining area, a 6-axis articulated arm robot of the type mantec r3 was used. The machining area and the articulated arm robot are surrounded with a sound protection of the size 3 m x 3 m x 3 m and connected to an exhaust removal with a filter for particles. To accelerate the blast medium, depending on the application, a round nozzle with an effective diameter of 13 mm and a newly developed flat nozzle with an effective area of 76 mm x 2 mm of Linde, Munich/Germany, were used. The entire dry ice blasting test stand is displayed in.

As blast medium cylindrical dry ice pellets from Linde, Leuna/Germany, were used. The diameter of the pellets was 3 mm, their length 2.0 to 18.5 mm at a mean length of 6.7 mm, their bulk weight 817.5 kg/m³ and their density 1,100 kg/m³.

3.2 Cleaning

The experimental investigation, the paint removal of metallic parts, the removal of silicone seals as well as the paint removal and cleaning of mounted printed-circuit boards by means of dry ice blasting were examined. The coated metallic parts are sheet steel made of St1203 with a workpiece thickness of 1 mm which are coated with stove enamelling paint Finadur 781 of Mankiewicz, Hamburg/Germany. The tested layer thickness is of 15 to 70 µm. The mounted printed-circuit boards are coated with conformal coatings on the basis of epoxy, polyurethane, and silicone resin. The epoxy coating of the type SL 1305-25 AQ has a layer thickness of 10 µm and the polyurethane coating SL 1308 FLZ of 50 µm. Both were produced by Lackwerke Peters, Kempen/Germany. The experimental investigations for removing from silicone seals were executed at components from an aluminium magnesium alloy of the A-Klasse of DaimlerChrysler. Further an application to dry ice blasting is the tube cleaning.

During the testing, the setting parameters pressure, dry ice mass flow, working distance from the blast nozzles to the blast material, jet angle of the dry ice free jet, blasting time, and advance rate of the blast nozzle varied.

For the evaluation of the tests, the removal width b and the therefore the removal rate P were measured with a measuring microscope PJN 322 of the company Mitutoyo, Tokio/Japan. The measuring of the roughness and of the surface profile of the processed surfaces was carried out with a PC-controlled electrical brush analyser Form Talysurf 120 L of Taylor-Hobson, Leicester/United Kingdom. To evaluate the surface topography and the structure of the material, a scanning electron microscope DSM 950 of the company Zeiss, Oberkochen/Germany was used.

3.2.1 Removal of Paint from Metal Components
One important application of dry ice blasting is the removal of paint from metal components. For this purpose, the parameters for processing sheet steel material were optimised. The aim was to find the optimum settings for stove enamelling paint with different coat thickness' between $s_L = 15$ to 70 µm. **Figure 3** shows the maximum advance rate v versus working distance a and jet angle ß as well as the removal rate P versus advance rate v for different coat thickness' s_L of stove enamelling paint for paint removal with dry ice blasting working with a round and a flat nozzle.

Using a round nozzle, the maximum advance rates for stove enamelling paint are in the order of v = 1.0 and 4.0 m/min, for the flat nozzle between v = 0.2 and 0.7 m/min. The highest advance rates were obtained with a jet angle of ß = 90 ° and a working distance of a = 150 mm for the round nozzle and of a = 50 mm for the flat nozzle.

The reason for the increase in the maximum advance rate with increasing jet angles is the increase in kinetic energy applied to the paint. Furthermore, the resultant thermal stress leads to cracking. This means increased pellet penetration into the cracks so that the paint is removed by the expansion effect.

With an increasing feed speed, the removal rate increases to a maximum value. Increasing the feed speed even more results in a strong decrease of the removal rate. The maximum removal rate value is not to be found in the very low or very high advance rate range. Instead, it can be found where the ratio between removal rate and feed speed is at its smallest. In the case of stove enamelling paint the removal rate decreases with an increase in coat thickness and constant feed speed.

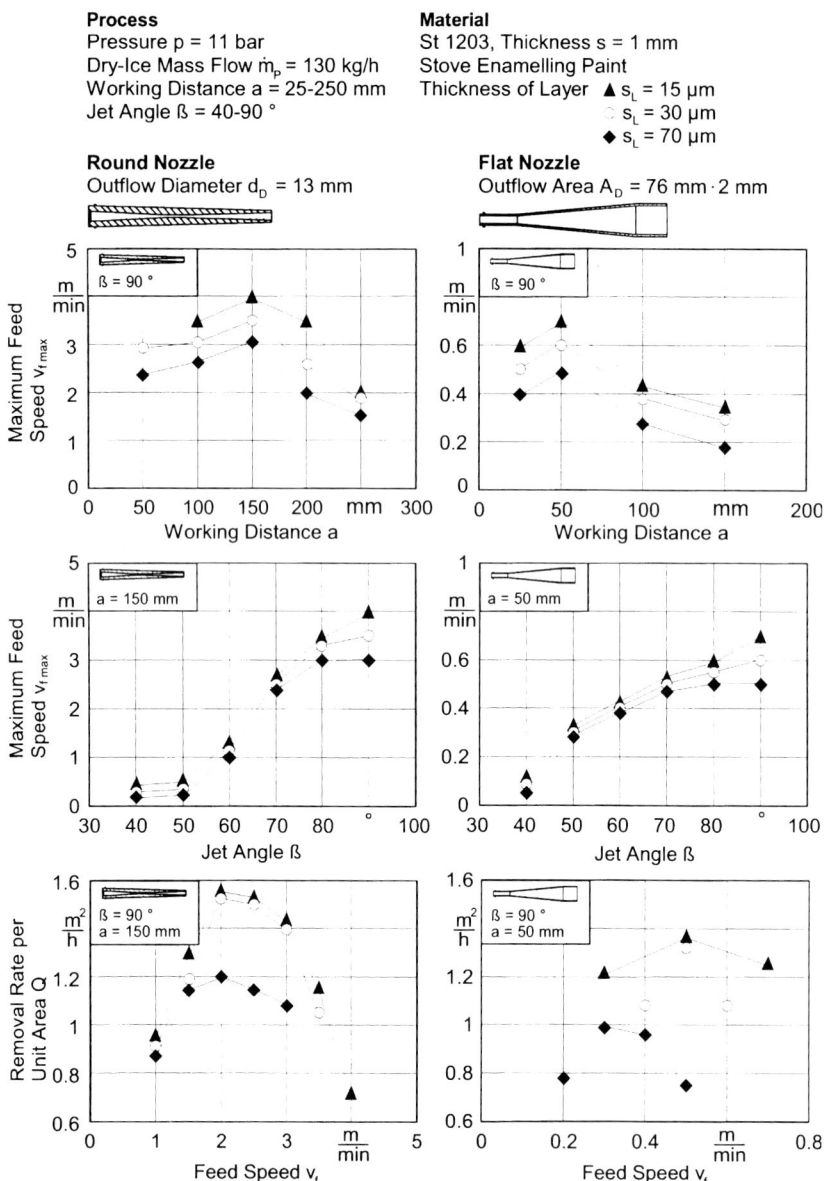

Figure 3: Maximum feed speed versus working distance and jet angle as well as removal rate versus feed speed for different thickness of layer of stove enamelling paint for paint removal from sheet steel material with dry ice blasting working with a round and a flat nozzle

D002/019 © IMechE 2002

3.2.2 Decoating and Cleaning of Mounted Printed-Circuit Boards

A new application of dry ice blasting is the decoating and cleaning of mounted printed-circuit boards. Objective is here the reuse of components and printed-circuit boards. The process is applied for the removal of conformal coatings within repair and recycling before the desoldering process of damaged or expensive components. With regard to dry ice blasting, the jet pressure will be reduced to 5 bars and the dry ice mass flow to 80 kg/h. **Figure 4** shows the paint removal rate P versus advance rate v for the removal of epoxy, polyurethane, and silicone coatings from mounted printed-circuit boards by dry ice blasting working with a round and a flat nozzle.

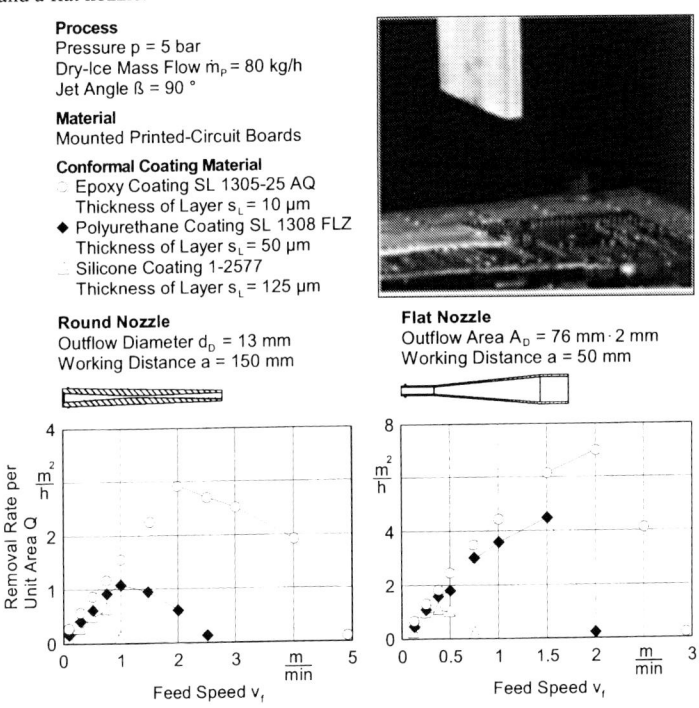

Process
Pressure p = 5 bar
Dry-Ice Mass Flow $\dot{m}_P$ = 80 kg/h
Jet Angle ß = 90 °

Material
Mounted Printed-Circuit Boards

Conformal Coating Material
○ Epoxy Coating SL 1305-25 AQ
   Thickness of Layer s_L = 10 μm
◆ Polyurethane Coating SL 1308 FLZ
   Thickness of Layer s_L = 50 μm
⸋ Silicone Coating 1-2577
   Thickness of Layer s_L = 125 μm

Round Nozzle
Outflow Diameter d_D = 13 mm
Working Distance a = 150 mm

Flat Nozzle
Outflow Area A_D = 76 mm · 2 mm
Working Distance a = 50 mm

Figure 4: Removal rate versus feed speed for the removal of epoxy, polyurethane, and silicone coatings from a mounted printed-circuit board by dry ice blasting working with a round and a flat nozzle

With an increasing advance rate, the removal rate increases to a maximum value. Increasing the velocity even more results in a strong decrease of the removal rate. The maximum removal value is not to be found in the very low or very high advance rate range. Instead, it can be found where the ratio between removal rate and advance rate is at its smallest.

Using a round nozzle, the maximum removal rate for the epoxy coating with a typical layer thickness of 10 μm is P = 3 m²/h, for the flat nozzle P = 7 m²/h. The highest removal rate was obtained with an advance rate of v = 2 m/min for the round nozzle and the flat nozzle. For the round nozzle the maximum removal rates for the polyurethane coating with a layer thickness of 50 µm and of the silicone coating with a layer thickness of 125 µm are P = 1 m²/h and P = 0.7 m²/h. The optimum advance rates are v = 1 m/min for the polyurethane coating and v = 0.7 m/min for the silicone coating. For the flat nozzle a removal rate of more than P = 4 m²/h with an optimum advance rate of v = 1.5 m/min for the polyurethane coating was obtained. The removal rate for the silicone coating was P = 1 m²/h with an advance rate of v = 0.4 m/min. The experimental results show, that with dry ice blasting it is possible to remove different types of conformal coatings at different layer thickness'. The obtained advance rates are compared to the removal of paint from sheet steel material very high.

3.2.3 Removal of Silicon Seals

The Automobile Industry uses lightweight construction alloys in the present models as engine material, e.g. aluminum magnesium alloys. The seal between the components is provided with silicone. With the repair and in the maintenance two problems occur by the silicone. Decomposition of the components by the bonding effect of the silicone and for removing the seal remainders is made more difficult, therefore existing at present no satisfying technology. At present the silicone seals are removed mechanically by manual scraping off. This is very time-consuming and very costly. A solvent or a chemical cleaning agent, with which the silicone can be solve, is not known. The silicone can be not completely removed by jets with sand, glass or steel due to elastics characteristics. Moreover these procedures lead a damage of the components. In addition is the geometry of the components very complex and that most several creating areas occur per component

The experimental investigations for removing from silicone seals were executed at components from an aluminium magnesium alloy of the A-Klasse of DaimlerChrysler **Figure 5**. In the context of parameter optimisation the mass flow of dry ice, the jet printing, the work distance of the angles between nozzle and component (free angle) as well as the processing speed were varied. It was worked with a high performance nuzzle with a outlet diameter of 14 mms and a length of 340 mm as well as a beam hose for transporting of the dry ice pellets with a length of 8 m. The experimental optimisation yielded a mass flow of dry ice from 130 kg/ h, a jet printing of 11 bar and a work distance of 100 mm at a free angle of 90 °. Depending of thickness and adhesion of the silicone residues as well as the breadth, deep and accessibility of seal joint could be obtained by there use of these parameters processing speeds dependently of up to 1 m/ min by complete cleaning of the creating area.

To the assessment of a possible material impairment through the dry ice blasting the surfaces and outer zones of the materials after the processing were analysed. This was occurring with the help of a surface measuring laser and a scanning electron microscope. No significant modifications of the surface roughness and the material structure could be proven at the processing speeds, which are relevant for removing the silicone seals. The microscopic recordings showed only by a punctual load of the material of more than one minute a abrasive effect. For the assessment of the material damages investigations at aluminium forgeable alloy AlMg3 were executed. Therewith the optimised process with parameters of removing from silicone seals were operated, however without processing speed.

D002/019 © IMechE 2002

It was shown, that removing of the silicone seals with dry ice blasting is technically feasible and cost savings are expect compared to present technologies. The renewed assembly of the engines after the processing is possible.

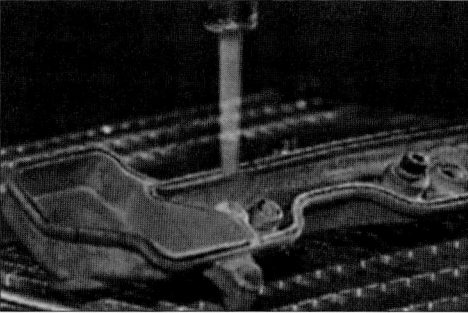

Figure 5: Removal of silicone seals by dry ice blasting from a valve cover of the A-Klasse from DaimlerChrysler

3.2.4 Tube cleaning

Germany accounts for more than 1 million km of sewage and waste water tubes and canals. Approximately 400,000 km tubes exist for fresh water and gas. Furthermore the industrial tube system has a length of several 100,000 km. These are mainly tubes in the chemical, petrol-chemical, pharmaceutical and the power supplying industry, respectively.

All tube systems are subject to pollution or clogging. The extent to which the tubes are polluted largely depends on the service conditions such as the quantity and quality of the flow, the installation, the age of the tubes and of course the flow medium. Cleaning or even new installation becomes necessary and unavoidable. The costs for the redevelopment of the public sewage system in Germany alone were estimated at 100 billion Euro. The demand for tube cleaning technologies and the growing market share resulted in the development of different cleaning processes and systems.

Today, the most relevant tube cleaning systems are based on mechanical, high-pressure, chemical or combined principles.

In mechanical tube cleaning different cutting heads are driven through the tubes. Drilling and milling tools or other rotating devices are the most common. Another possibility is mechanical cleaning in process. Solid bodies flow within the medium. The tube is cleaned by the friction between the tube wall and the solid bodies.

In the field of high-pressure tube cleaning pressurised air or water is pumped through the tubes. Special abrasives can be added to deepen the impact and the cleaning effect.

In chemical tube cleaning different chemical solvents are used depending on the pollution. These can be acids, alkaline solutions, hydrocarbons or aqueous solutions.

All of these systems have in common that they are only applicable for specific mains network systems. All may show poor cleaning results, are limited by the tube installation and bendings, affect or even damage the tube surface, are expensive or ecological critical.

Due to the limitations of conventional tube cleaning systems the Institute for Machine Tools and Factory Management and two industrial project partners focus on the development of an innovative, flexible and eco-efficient cleaning system.

Especially for tube cleaning the dry ice blasting technology has considerable advantages via conventional systems:

- No blast medium residue.
  Due to the sublimation of the carbon dioxide at the tube walls no blast medium has to be removed and disposed.
- Non-polluting.
  Only the decoated pollution has to be removed from the tube and disposed. No critical chemicals were used.
- No abrasive damage.
  The hardness of dry ice pellets is too little to damage the tube material.
- No corrosion or drying.
  In the process no water or other chemicals are used.
- No expensive assembly or disassembly.
  The cleaning is possible without demounting of the tubes.-
- No electricity.
  The process is completely pneumatic, therefore it is mobile and flexible.

The aim of the project is the development of a prototype tube cleaning system using the dry ice blasting technology. The project is performed with two partners. *COMBAS, Rohrsanierungs-, Vertriebs- und Dienstleistungsgesellschaft mbH*, a tube cleaning and redeveloping company from Brandenburg/Germany and *Green Tech Trockeneisstrahlanlagen GmbH & Co. KG*, a manufacturer of dry ice blasting devices from Bavaria/Germany. The project is promoted by the *Bundesministerium für Wirtschaft und Technologie* (Ministry of Commerce and Technology) for a period of 21 months.

Due to the diversity of materials, pollutions, diameters, installations and conditions of tubes it is not realistic to develop a universal tube cleaning system reaching constant cleaning results. Regarding the possibilities of the dry ice blasting technology the project partners stated some basic requirements for the new system:

- The minimum cleaning length shall be 100 m. Shorter section length would not be economical due to too many revision ducts.
- 300 mm was chosen as a representative tube diameter. An adaptation to other diameters will follow.
- Tubes and canals to be cleaned shall have a circular profile. An adaptation to other shapes such as a rectangular profile e.g. in ventilation and air conditioning canals will follow.
- The system shall be able to drive through tube bendings with small radii in relation to the tube diameter.
- The sound pressure level outside the tube shall meet residential area limitations.

In cooperation with the project partners, the basic technological concept for the prototype tube cleaning system was developed. **Figure 6** represents the concept of the tube cleaning system with dry ice blasting.

A pressure blasting, one hose system directs the dry ice through a flexible hose line to the respective tubes. The dry ice then splits into two jets and hits the tube wall at an impact angle of 90°.

Experiments have shown that other possibilities to direct the dry ice jet onto the tube walls such as a circular widening of the jet do not reach the necessary force to remove resistant pollutions. The jet splitter is rotatable to direct the jets onto the whole circumference of the tube. The rotation is accomplished by an air-motor with an adjustable rotation speed. The rotation speed is a function of the transport speed.

D002/019 © IMechE 2002

The feed is realized by a return-motion installation. Such a device enables a feed forward and backward, respectively. It is also possible to transport the hose independent of its length and weight.

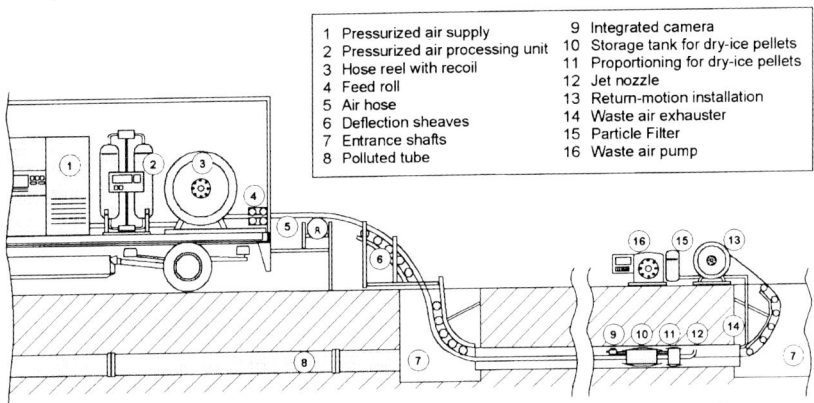

1 Pressurized air supply	9 Integrated camera
2 Pressurized air processing unit	10 Storage tank for dry-ice pellets
3 Hose reel with recoil	11 Proportioning for dry-ice pellets
4 Feed roll	12 Jet nozzle
5 Air hose	13 Return-motion installation
6 Deflection sheaves	14 Waste air exhauster
7 Entrance shafts	15 Particle Filter
8 Polluted tube	16 Waste air pump

Figure 6: Basic concept of the tube cleaning system with dry ice blasting

The parameter optimisation of the process will be performed at the Institute for Machine Tools and Factory Management. Therefore an experimental test-bed for tube cleaning was installed. The system will be tested on Perspex as well as on real tube sections up to a length of 4 m. The cleaning results and the impact on the tube materials will be analysed. The results of the analysis serve as a basis for the parameter optimisation e.g. jet pressure and for nozzles to be added on the jet splitter outlets as well. The shape of the changeable nozzles largely depends on the pollution.

3.3 Surface Preparation

For an optimal adhesion of adhesives with aluminium components, how they are used for example in the automobile industry, a chemical activation or a physical modification of the surfaces is necessary. However industrially applicable procedures for the surface modification are further looked for as well as the appropriate processing parameters are empirically determined. Previous investigations showed that by degreasing with solvents, pickling with chrome sulphuric acid, blasting with corundum, flaming and different plasma processing an adhesive strength increase of adhesive bindings with aluminium plates can be obtained.

In the context of an experimental study also dry ice blasting for this application was examined. For the determination of the adhesive strength of the adhesive bindings rehearses of the aluminium forgeable alloy AlMg3 according to DIN EN 1645 were analysed. It was shown that surface roughness and - topography have a substantial influence on the adhesive strength of the adhesive bindings. Herewith a higher surface roughness and a surface topography with undercut affect positively on the adhesive strength.

Dry ice blasting can be applied for the direct surface modification by aluminium forgeable alloys [17]. Defined surface roughness can be adjusted **Figure 7**. Herewith through dry ice blasting corrosion and pollution layers are removed and the surface is increased due to larger roughness. Additionally grain surfaces at the surface are exposed, which thus activate the surface as well as produce defined surface textures with undercut, small hollows and impact crater. The elaborated technological database covers the connections between the adjusting parameters feed speed, work distance, jet impact angle, jet printing and mass flow at dry ice as well as the surface and outer zone characteristics.

The automation of the dry ice blasting to the specific pre-treatment of binding areas is possible. Next to the use to the strength increase of binding joint, the procedure can be used also for the pre-treatment by components before coating processes.

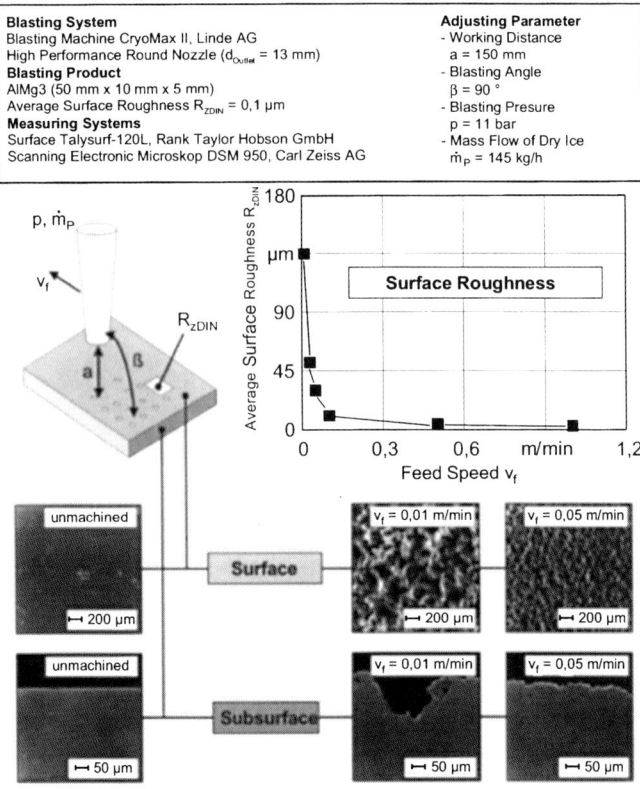

Blasting System	Adjusting Parameter
Blasting Machine CryoMax II, Linde AG	- Working Distance
High Performance Round Nozzle (d_{Outlet} = 13 mm)	a = 150 mm
Blasting Product	- Blasting Angle
AlMg3 (50 mm x 10 mm x 5 mm)	β = 90 °
Average Surface Roughness R_{zDIN} = 0,1 μm	- Blasting Presure
Measuring Systems	p = 11 bar
Surface Talysurf-120L, Rank Taylor Hobson GmbH	- Mass Flow of Dry Ice
Scanning Electronic Microskop DSM 950, Carl Zeiss AG	$\dot{m}_P$ = 145 kg/h

Figure 7: Average surface roughness versus feed speed as well pictures of surface and subsurface

In the context of further investigations the hardening of the material surface, the hardening of the surface outer zones as well as internal stress conditions in the area near of the surface were determined. Herewith the metals X5CrNi18.9, CuZn37 and AlMg3 with dry ice blasting were machined. For example the Vickers-hardness of the heat treatable steel X5CrNi18.9 can be increased with a feed speed by 0,01 m/min from 200 HV to 335 HV. This hardness increase could be proven into a depth of 140 μm. The roughness and the internal stress of the heat treatable steel increases thereby only slightly. In case of the brass alloy CuZn37 the hardness

D002/019 © IMechE 2002

can be increased only over up to 20 %, however the printing internal stress rises. For the aluminium forgeable alloy AlMg3 no significant hardness increase can be obtained. The material possesses small output hardness, therefore it comes with dry ice blasting to an brittleness and plastic deformation at the surface. This has merely only a higher surface roughness.

3.4 Cutting

In the context of a feasibility study with the GFM GmbH Austria, Steyr/Austria, the application of the dry ice blasting to jet cuts by composites was examined. Herewith a new process engineering for the production of the blast medium was tested, cut-specific jet nuzzles was developed as well as a process optimisation was executed. The composites consist of glass fibre panels with honeycomb core, Honeycomb structures, are applied in the air and space travel industry, in the automobile industry and in the sport instrument industry.

The results show that the production of harder dry ice particles is technically possible and this has an increase of the abrasive effect. The developed jet nuzzles with effective diameters from 3 to 4 mm have a focused open jet and high jet output speeds, however the blast medium must be cut up with a crusher on particle diameters from 100 to 200 µm. In the context of a process optimisation the influence of the jet distance and the cutting speed on the cut joint quality was examined. It could be stated that the examined composite materials can be cut with dry ice blasting. With material thickness of 8 mm cut joint breadths of 3.5 mm with cutting speeds by 0,1 m/min were obtained.

In comparison with the cutting technologies that are used nowadays, the cutting with dry ice is even not competitively due to the low cutting speed. However through the sublimation of the jet medium, the dry processing and cold processing as well as the handling flexibility it offers advantages, which are interesting for future applications

4 CONCLUSION

The executed feasibility studies have shown, that dry ice blasting is a flexible usable and environmentally beneficial manufacturing process. The applications extend from cleaning, decoating, hardening and pre-treating up to the trimming and cutting of materials. Particularly in the cleaning a series of companies apply this process successfully. In the next years is to be calculated along with furthermore a strong increasing stock exchange for the dry ice blasting. Insiders of the business estimate conservatively that the dry ice cleaning market will grow for approximately 25 % per year in the next 5 years.

REFERENCES

1   **Harms, D.; Seyd, G.** and **Bergbauer, A.** Verfahren zur Reinigung von metallischen Oberflächen, insbesondere radioaktiv verseuchten Oberflächen. Patent DE 2058766, Siemens AG, Berlin/Germany, Deutsches Patentamt, registration date: 30/11/1970.

2   **Rice, E.** and **Franklin, C.** Method for the removal of unwanted portions of an article. Patent US 3,676,963, Chemotronics International Inc., Michigan/USA, United States Patent Office, registration date: 08/03/1971.

3   **Rice, E.** and **Franklin, C.** Method for the removal of unwanted portions of an article by spraying with high velocity dry ice particles. Patent US 3,702,519, Chemotronics International Inc., Michigan/USA, United States Patent Office, registration date: 12/07/1971.

4   **Fong, C.** Sandblasting with pellets of material capable of sublimation. Patent US 4,038,786, Lockheed Aircraft Corp., California/USA, United States Patent Office, registration date: 27/08/1975.

5   **Uhlmann, E.; Axmann, B.** and **Elbing, F.** Stoßkraftmessung beim Strahlen mit CO_2-Pellets, Zeitschrift für wirtschaftlichen Fabrikbetrieb ZWF, 1998, Vol. 93, No. 6, p. 240-243.

6   **Uhlmann, E.; Axmann, B.** and **Elbing, F.** Trockeneisstrahlen - neue Erkenntnisse durch Hochgeschwindigkeits-Videographie, Journal für Oberflächentechnik JOT, 1998, Vol. 38, No. 7, p. 46-51.

7   **Uhlmann, E.; Axmann, B.** and **Elbing, F.** Reinigen mit Trockeneisstrahlen in der Austauschmotorenfertigung, VDI-Z, 1998, Vol. 140, No. 9, p. 70-72.

8   **Axmann, B.** and **Elbing, F.** Strahlverfahren. Zeitschrift für wirtschaftlichen Fabrikbetrieb, 1997, Vol. 92, No. 3, p. 76-77.

9   **Axmann, B.** and **Elbing, F.** Kompetenzzentrum Strahlverfahren. Zeitschrift für wirtschaftlichen Fabrikbetrieb ZWF, 1998, Vol. 93, No. 3, p. 57.

10   **Uhlmann, E.** and **Elbing, F.** Reinigen und Entschichten mit CO_2-Trockeneisstrahlen. Schwerpunktseminar Reinigen/Abtragen, Institut für Werkstoffkunde, Universität Hannover, Hannover/Germany, 02/03/1998.

11   **Uhlmann, E.; Axmann, B.** and **Elbing, F.** Trockeneisstrahlen - neue Erkenntnisse durch Hochgeschwindigkeits-Videographie. Journal für Oberflächentechnik JOT, 1998, Vol. 38, No. 7, p. 46-51.

12   **Uhlmann, E.; Axmann, B.** and **Elbing, F.** Reinigen mit Trockeneisstrahlen in der Austauschmotorenfertigung. VDI-Z, 1998, Vol. 140, No. 9, p. 70-72.

13 Spur, G.; Uhlmann, E. and Elbing, F. Experimental Research on Cleaning with Dry ice Blasting. Proceedings of the International Conference on Water Jet Machining WJM `98, Krakow/Poland, 06/11/98, p. 37-45.

14 Uhlmann, E.; Axmann, B. and Elbing, F. Cleaning and Decoating in Disassembly. Proceedings of the 4th World Congress R `99 - Recovery, Recycling, Re-integration. Geneva, Schwitzerland, 02-05/02/1999.

15 Spur, G.; Axmann, B. and Elbing, F. Model for Kerf Profile and Jet Velocity in Abrasive Water Jets. Proceedings of the 5th Pacific Rim International Conference on Water Jet Technology, New Delhi/India, 1998, p. 407-420.

16 Uhlmann, E.; Axmann, B. and Elbing, F. Stoßkraftmessung beim Strahlen mit CO_2-Pellets. Zeitschrift für wirtschaftlichen Fabrikbetrieb ZWF, 1998, Vol. 93, No. 6, p. 240-243.

17 Uhlmann, E.; Spur, G. and Elbing, F. Trockeneisstrahlen: Oberflächen- und Randzonencharakterisierung von AlMg3. Proceedings of the International Conference Manufacturing `01, Posen /Poland, 08-09/11/2001, p. 457-466

Innovative machining technologies and tools for the disassembly of consumer goods

E UHLMANN, G SPUR, F ELBING, and **J DITTBERNER**
Institute for Machine Tools and Factory Management, Technical University of Berlin, Germany

ABSTRACT

Ecologically harmless disposal of used technical consumer products will become mandatory for their producers, as well as for importing companies. This disposal policy will focus on product and material loops: Used products will be disassembled and the parts and materials then recycled. Due to environmental and legislative reasons the importance of disassembly as an important step in the process of recycling is steadily rising. The paper presents developed technologies and tools for the disassembly of consumer products. An automatic pilot disassembly system is realised for testing the disassembly of washing machines. The aim is to recover materials and re-use components. Different destructive processes such as plasma arc cutting, water jet cutting, drilling and abrasive cutting were optimised to disassemble washing machines, especially to dismantle the cases. A new cryogenic cutting tool will be developed for the disassembly of metal and plastic parts.

1 INDUSTRIAL DISASSEMBLY

Not long ago, a product life started with assembly and ended with disposal. Today, in times of increasing waste quantity and decreasing natural resources, life cycles are invented in which materials are re-processed and components are re-used [1]. Old technical consumer products are disassembled. Disassembly is a major operation in the recycling of used technical products. But disassembly is not as highly automated as assembly yet. The standard of automation for disassembling is still very low and therefore dismantling is often done manually [2, 3]. Employment is slow and cost-intensive, especially in an highly industrialized part of the world as Europe. Hence recycling is more economic by automated disassembly. On the other hand, the human worker is the most flexible tool available. In the face of the high variety of types of a technical product and the state they are in after use, it is clear that many different information have to be respected when choosing disassembly tools, processes and strategies.
The conclusion is, that flexibility and rapidness are two main requirements for industrial disassembly. At the Technical University Berlin a Collaborative Research Centre works on basics, tools and processes for disassembly. The objective is to increase the effectiveness of disassembly through optimised disassembly techniques. The paper presents a general introduction of the four project areas and lies a focus at the examinations of a sub-project dealing with the separation of insoluble connections.

1.1 Collaborative Research Centre 281

At the Technical University Berlin and the Berlin University of Arts, the Collaborative Research Centre 281 „Disassembly Factories" promoted by the Deutsche Forschungsgemeinschaft (DFG) investigates industrial disassembly. The project was started in 1995 and is in the third promoted research period now. The centre is divided into four different project areas which deal with different aspects of disassembly [4]. Project area A "Tools and Processes" invents and optimises single disassembly techniques and tools as well as a pilot disassembly system. Project area B deals with the logistic problems of collection an distribution in disassembly and describes ways to implement the disassembly factory in the environment. Project area C determines the value of a disassembly product and plans disassembly. Project area D designs disassembly-friendly products and facilitates disassembly.

1.1.1 Tools and Processes

In order to control and economically optimise disassembly processes, a basic knowledge of these processes is required. Project area A delivers these elementary processes as well as tools, production facilities, information and sensor systems for disassembly [4]. Within the research period 2001 to 2003 the goal of project area A is to develop highly flexible, adaptable and modular process and tool concepts for different life cycle scenarios. The research on cutting processes, information, sensor and disassembly systems has to be combined and verified within a pilot disassembly system. The focus is on advantages, disadvantages and restrictions of processes and tools concerning flexibility, modularity and adaptability.

Challenges for further investigations of disassembly processes and tools include fast, safe, and economical disassembly of structural parts and materials. As an example the separation of washing machine housings is tested in order to achieve fast access to valuable components.

New modular tools for manual and automated use are developed to disconnect joints and to handle disassembly objects. This is realised by the generation of new acting surfaces. Tools, developed during the first and second research period, are integrated into a modular construction kit as functional units.

Further process and tool developments, e. g. desoldering, are related to non-destructive disassemblies in order to preserve component quality for reuse. A clamping technology concept for disassembly combines clamping and handling of different products. Within the project area, a cleaning technology with laser and dry ice blasting re-use and further use of disassembly products is developed.

The Life Cycle Unit (LCU) collects life cycle related data in order to support disassembly processes. Detailed information about product and component quality is available throughout the entire life cycle. Product information about the steering of disassembly and re-assembly are equally gathered.

The development of processes and tools, security directives, information systems, effectors and sensors is the first step in the direction of automation. The product area is divided into six sub-projects which deal with separating of insoluble connections, shape independent end-effectors for non-destructive disassembly, flexible clamping technology, protection of labour and safety precautions, cleaning technologies, sensor guided processes, and the realization of a pilot disassembly system.

D002/020 © IMechE 2002

1.1.2 Logistics and Urban Development

It is necessary to integrate logistics, disassembly factories and urban structures in order to achieve closed product and material loops. The project area B develops logistical disassembly systems in close relation to the pilot-disassembly system of product area A, as well as other in-house logistic systems. Object is to match external logistics with the logistics of customers as well as with suppliers. This requires a certain organization which has to be designed and installed prior to realization [4].

The project is divided into three sub-projects which deal with the process oriented development of logistical systems, the creation of a regional network of disassembly factories and the interaction between disassembly factory and city.

1.1.3 Product Evaluation and Disassembly Planning

A substantial requirement of disassembly is the investigation of the individual product. Planning and realization of disassembly facilities is strongly influenced by uncertainties which result from usage of products. To meet resulting requirements the systems have to be flexible and offer an online control potential reaching far further than any given system available today.

Information relevant for disassembly, such as experience gained from previous efforts, general product information, and the usage related condition of the product, must be considered together. The disassembly process of an individual product cannot be precisely predicted ahead of time and must be planned just prior to the start of the process or even during the process. This situation results from different conditions in which a product may be dependent on the influence of usage.

Evaluation of disassembly relies heavily on the difference between costs and benefits. On the side of costs it is most important to give value to a wide range of different items like emissions, resource consumption or social costs. To evaluate things as social costs, it becomes necessary to find appropriate cost accounting systems. As one result, the system found can help improve existing products and, additionally, improve the quality of new products.

The project area C is divided into three sub-projects, which deal with methods of assessing recycling suitability and computer aided disassembly planning and control as well as with economic and ecological course of action [4].

1.1.4 Product Design and Ease of Disassembly

Disassembly efficiency will be increased if design for ease of disassembly requirements such as reusable component architecture or pure material use are met. Emphasis lies on joints. It is demanded that joints not only offer accessibility for tools, but also that they are easy to disassemble. Deduction of design criteria provides a great deal of support for assessment during the design stage. To predict end-of-life product conditions, simulation tools are needed which require an information network that provides and receives information from producers, users and recyclers.

The project area D is divided into four sub-projects, which product with the selection, embodiment and arrangement of fasteners, with design guidelines for structures and joining components, with simulation tools for a disassembly-adequate product design, and with the disassembly-oriented information-technological infrastructure [4].

2 SEPARATION OF INSOLUBLE CONNECTIONS

Disassembly products consist of many different parts. The parts are hold together by joints. These joints are welded connections, screws or point forming connections. Often it is not possible to demount these joints. They have to be destroyed. The sub-project A1 deals with the separation of insoluble connections [4, 5]. Within it, processes are optimised for disassembly and compared with each other. Examined destructive processes are plasma arc cutting, drilling, water jet cutting and abrasive cutting. The processes are estimated and integrated in the pilot disassembly system. First example for a disassembly product was a washing machine. The main field to use destructive processes in the disassembly of washing machines is to open up the case to gain fast access to re-usable components.

Most cases of washing machines are made of steel, St1203 or similar. Some components are made of V2A, less of AlMg3. Coatings used are enamel paints, dry lacquers or two components lacquers. Thus combinations of those materials and coatings were used for the examinations. Following, the examinations on plasma arc cutting and water jet cutting are described. The chosen material thickness was 1 mm, 2 mm and 3 mm.

2.1 Plasma Arc Cutting

For the disassembly of washing machines, specific disassembly sequences were elaborated for each type. The washer case is hereby separated by means of plasma arc cutting in such a way that a frame remains, which the inner components are attached to. The aim of this disassembly step is to reach the inner components and assemblies as fast and cost-effective as possible. A robotic plasma arc cutting device PA-S 45 CNC by Kjellberg, Finsterwalde/Germany, was used. The installation works with a transmitted plasma arc at cutting amperages of 40 A to 130 A, a cutting voltage of 160 V and a cutting power of 20 kW. In the course of technical investigations, a qualitative gas analysis was carried out as well as a process optimisation with regard to maximum cutting speed and a completely cut through workpiece, minimum mechanical and thermal component damage and minimum removal particle mass [6].

The maximum cutting speed was obtained in experiments with the standard experimental set up by Tagucci, during which the parameters cutting amperage, cutting gas, workpiece material, and material thickness were varied. The cutting gases tested were air, Ar/H_2, and $Ar/H_2/N_2$. The maximum cutting speeds could be obtained with air. How fast the cut with air ever may be, there is a high emission of removal particles. With increasing cutting amperage the maximum cutting speed increases too. At a cutting average of 125 A a maximum cutting speed of 12 m/min can be achieved. The cutting amperage and the cutting gas with a minimum removal particle mass were determined. **Figure 1** displays the removal particle mass as a function of the cutting amperage in the case of plasma arc cutting of a workpiece similar to the washer case of the type Bosch V454. After an optimisation the particle emission during plasma arc cutting of a washer case could be reduced by 95 % compared to the initial setting, which was a cutting amperage of 125 A and the cutting gas air. The conclusion is that the removal particle mass is almost irrespective of the cutting speed in the case of a defined separating cut and that it is only minimal when cutting gas Ar/H_2 and a cutting amperage of 75 A are applied. A disassembly time of three minutes is calculated for the industrial disassembly of a complete washing machine case, if optimised parameters are applied during the separation of the four lateral walls of the washer case. After that, the assemblies and components can be disassembled in a destruction-free manner.

D002/020 © IMechE 2002

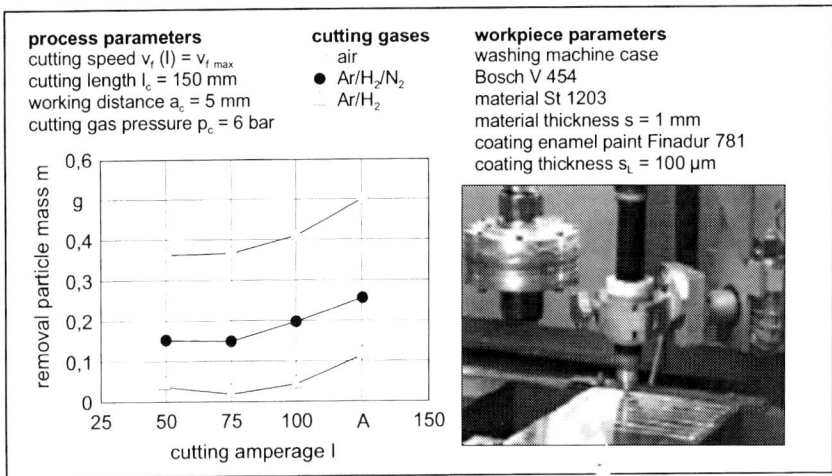

process parameters	cutting gases	workpiece parameters
cutting speed v_f (l) = $v_{f\ max}$	air	washing machine case
cutting length l_c = 150 mm	● Ar/H$_2$/N$_2$	Bosch V 454
working distance a_c = 5 mm	Ar/H$_2$	material St 1203
cutting gas pressure p_c = 6 bar		material thickness s = 1 mm
		coating enamel paint Finadur 781
		coating thickness s_L = 100 μm

Figure 1: Removal particle mass as a function of cutting ampe 'age and cutting gas during plasma arc cutting of a washing machine case

Including the results of the above described examinations the construction of the plasma arc plant and the periphery were altered and integrated into a hybrid pilot disassembly system. Due to the high cutting speed, good automation and low cutting costs, plasma arc cutting was used in the pilot disassembly system to open washing machine cases from various suppliers. Further examinations take place on cutting poles and wires by plasma arc cutting.

2.2 Water Jet Cutting

A robotic abrasive water jet cutting device from Ingersoll Rand, Bad Nauheim/Germany, was used for the examination of water jet cutting in disassembly. Again metallic materials were processed. The maximum working pressure was 320 MPa, the average abrasive mass flow 350 g/min and an abrasive garnet mesh 80 was used.

First focus lies on the optimisation of the maximum traverse rate where a workpiece is cut through. Dependent from the material and the workpiece thickness a model was build to predict the maximum traverse rate. Beside the material properties, the working distance, abrasive mass flow and the working pressure quantifies the traverse rate. It is highest at maximum pressure and an abrasive mass flow of 350 g/min [7].

Another task of the investigations on water jet cutting was to determine and explain the jet properties, particularly the influence of the secondary jet on inner components and assemblies. If water jet cutting is used for three-dimensional machining, damages occur caused by the primary and the secondary jet. The primary jet is the jet prior to machining and the secondary jet exists behind the workpiece that is machined and dies in infinity or when striking inner components. Within disassembly the chances are high, that the secondary jet strikes an inner component or other assemblies. When the so called secondary jet distance, the distance between the primary workpiece and inner components, is low, re-usable parts are in risk of being damaged. Thus the objective is to minimize the energy of the secondary jet and to predict its influence on other components. The secondary distance is measured on a straight line, which runs vertically to the primary work piece through the jet exit point.

In analogy to the calculation of the kerf geometry, it is possible to calculate the depth of the damage for the secondary jet. Essentially, the damage of the inner components depends on the traverse rate and the secondary jet distance from the primary workpiece. When neglecting the variations in cutting rate, the damage inflicted by the secondary jet during cutting with maximum traverse rate is zero. Thus the traverse rate must be seen in relation to the maximum traverse of the primary work piece. The closer the traverse rate gets to the maximum traverse rate, the lower is the secondary jet energy and the possible damage. Yet, not only the value of the secondary jet changes, but also the direction of the jet given by the exit angle.

The damage by the secondary jet was carried out in analogy to the tests on the primary abrasive water jet. Thus, the secondary jet spreads according to the same rules as the primary jet. As a result, the damage of inner components can be minimised and predicted. With this knowledge, it is possible to set cutting strategies for disassembly, which avoid damages on re-usable components.

The abrasive water jet was integrated into the disassembly process of standard washing machines to open the plastic lye container and to separate it from the washing cylinder made from stainless steel.

Figure 2 shows the disassembly stages of a washing machine. It starts with the opening of the case, followed by the separation of the swinging system which is then stripped from weights, pump and motor. Afterwards the water jet opens up the lye container and the washing cylinder is removed.

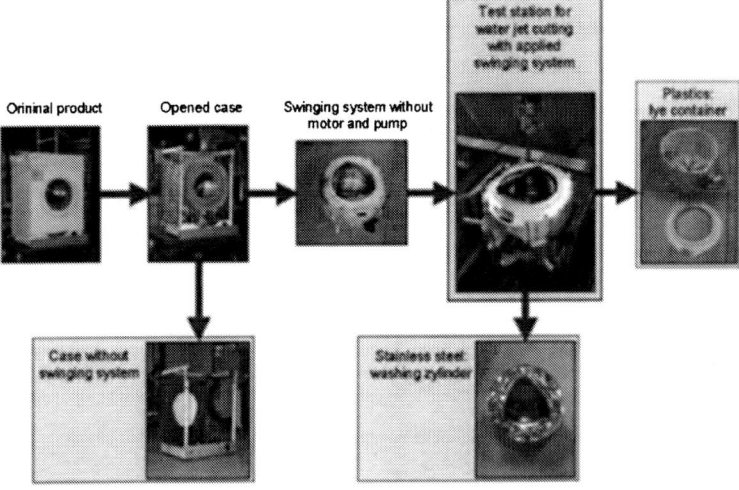

Figure 2: Disassembly stages of a washing machine

D002/020 © IMechE 2002

3 CONCLUSION

The challenges for industrial disassembly are new life-cycle management laws, a growing recycling and maintenance industry, the handling of contaminations and harmful chemicals in recyclable products, and the re-use of components from recyclable products. An approach to this field is made by the Collaborative Research Centre "Disassembly Factories". Focuses lie on the disassembly of consumer goods and on the cleaning of components for re-use. Object is to develop processes and tools for a fast and economical disassembly. The sub-project "Separation of Insoluble Connections" does a basic approach to this.

The described examinations show the independence of water jet cutting and plasma arc cutting from material and working direction. The processes were compared with other processes such as drilling, abrasive cutting and a self developed strike-cutting tool. Plasma arc cutting is the most flexible and fastest process for the disassembly of metallic consumer goods. Only when it comes to disassemble non-metallic components water jet cutting is even more effective, otherwise it is second choice. The flexibility of these processes is their main advantage. For specific tasks, other tools may have higher cutting rates or a lower risk of damaging inner components.

Therefore a combination of these processes with a cryogenic pre-processing is in development. By cooling material to a temperature of nearly minus 200 °C by liquid nitrogen it is possible to embrittle it. Thus less forces occur while cutting material and it is possible to increase the cutting speed.

REFERENCES

1 **Spath, D.; Klimmek, M.** and **Tritsch, Ch.** (1996) Technologies for efficient disassembling and dismantling of discarded technical products. Proceeding of the 3rd International Seminar on Life Cycle Engineering, Zurich/Switzerland.

2 **Scholz-Reiter, B.; Scharke, H.** and **Hucht, A.** (1999) Flexible robot-based disassembly cell for obsolete TV-sets and monitors. Robotics and Computer Integrated Manufacturing, Vol. 15, p. 247-255.

3 **Feldmann, K.** and **Meedt, O.** (1996) Recycling and disassembly of electronic devices. Life cycle modelling for innovative products and processes (Krause, F.-L., Jansen, H.), London, Chapman & Hall, p. 233-245.

4 **Seliger, G. et al.** (2000) Finanzierungsantrag 2001-2003. Sonderforschungsbereich 281 "Demontagefabriken zur Rückgewinnung von Ressourcen in Produkt- und Materialkreisläufen", TU Berlin.

5 **Uhlmann, E.; Axmann, B.** and **Elbing, F.** (1999) Fast disassembly of point forming and welded connections. Proceedings of the 4th World Congress R '99 - Recovery, Recycling, Re-integration, Geneva/Switzerland, p. 402-411.

6 **Spur, G.; Uhlmann, E.; Elbing, F.; Dittberner, J.; Sundaresan, S.** and **Thantry P, B.** (2000) Flexible automatic disassembly for the recycling of consumer goods. Advances in manufacturing Technology XIV, Proceedings of the 16th National Conference on Manufacturing Research, London,/England, p. 407-411.

7 **Uhlmann, E.; Axmann, B.** and **Elbing, F.** (1999) Model of kerf and simulation of damage in abrasive water jet cutting, Production Engineering.

Dry machining of cast aluminium automotive wheels – innovative cutting tool design for improved machining performance and environmentally conscious manufacturing

I S JAWAHIR, S CHEN, and D TROUTMAN
Center for Robotics and Manufacturing Systems, University of Kentucky, Lexington, Kentucky, USA
B ALLISON
Central Manufacturing Company, Paris, Kentucky, USA
G POULACHON
ENSAM, Labo Usinage, Cluny, France

Abstract

Environmentally conscious, coolant-free dry machining technologies have emerged in recent years as viable alternatives to the traditionally known machining processes involving flood-cooling methods. Contour turning of cast aluminum wheels in the automotive industry poses significant challenges in process and product innovation for improved machining performance, particularly with the pressing need for chip control for defect-free cast aluminum products. Thus a further study is required regarding the design and development of an innovative cutting tool, which is essential for facilitating the favorable chip flow leading to improved chip control. The anticipated broader functionality of the new tool design is also to facilitate a coolant-free dry machining process.

This paper will present the results of a recent analytical and experimental study involving the use of an analytical model for predicting the chip flow direction and comparison of predictions with experimentally observed chip flow using high speed filming techniques. The identification of "problem areas" during the contour turning of the wheel profile led to process innovations in machining and a new product development, with great economic impact and also promotes a change toward environmentally conscious machining. This paper presents the details of this process improvement and the development of a new, innovative cutting tool design to provide coolant-free dry machining with improved chip control in contour turning of cast aluminum wheels used in the automotive industry. Shop floor tests using the new cutting tool show significant improvements in chip flow and machining performance.

1. INTRODUCTION

Coolants and lubricants used in machining processes help to improve the machining performance measures such as cutting forces, tool-life/tool-wear, chip-form/chip breakability, surface roughness, part accuracy, etc. Although many technological advantages are associated with the use of coolants and lubricants, many hazards to the health of personnel and to the environment exist. The cost of acquiring, maintaining, and disposing of coolants and lubricants represents about 16% of total production cost, which is a significant portion at the total machining cost. Therefore, much motivation is present for industry to move toward dry machining (machining without the use of any cutting fluid).

In the automotive industry, the use of lightweight materials can help reduce vehicle weight and improve fuel economy. The tendency for reducing weight has given rise to a gradual decrease in the amount of steel and cast iron used in vehicles and a corresponding increase in the amount of alternative materials, especially aluminum and plastics. Aluminum usage in automotive applications has grown more than 80% in the past 5 years. A total of about 110 kg of aluminum was used per vehicle in 1996, and is predicted to rise to 250 or 340 kg, with or without taking the body panel or structure applications into account, by 2015 [1]. In North American, there is also a broad range of opportunities for employing aluminum in automotive powertrain, chassis, and body structures. Currently in North America 240,000 tons of aluminum products are used per year in the automobile industry [2].

Contour turning is widely used in many automotive industry applications. It gives rise to unique problems due to continuously changing cutting conditions, wide variations in chip flow and the consequent chip-form, and surface finish issues due to geometric variations along the length of cut. Contour turning of cast aluminum wheels in the automotive industry poses significant challenges in process and product innovation for improved machining performance, particularly with the pressing need for chip control for defect-free cast aluminum products. This paper presents a methodology for the development of an innovative cutting tool, which is essential for facilitating the favorable chip flow leading to improved chip control. The anticipated broader functionality of the new tool design is also to offer a coolant-free dry machining process.

2. PREDICTIVE MODELING OF CHIP FLOW IN CONTOUR MACHINING

1.1 Review of Predictive Models for Cutting Forces and Chip Flow
Considerable work has been done on modeling of cutting forces and chip flow in machining operations. Early work by Colwell [3] assumed that the chip flow over the rake face of the cutting tool was perpendicular to the major axis of the projected area of cut (for nose radius tool, this is the segment joining the extremities of the engaged cutting edge). However, the effect of the rake and inclination angles was included as an empirical constant with no relation to the cut geometry. Stabler [4] introduced the relation that the chip-flow angle will equal the inclination angle (Stabler's flow rule). Later, he modified his relation by introducing a material dependent constant without considering other tool geometry parameters [5]. Brown and Armarageo [6] investigated the mechanics of chip flow in oblique machining. Armarego [7] utilized a generalized mechanics of cutting approach for prediction of cutting forces and chip flow in machining operations. In a more recent work [8], the model has been updated

with some new geometric features through elaborate experimental work. Lin et al. [9], Young et al. [10], Arsecularatne et al. [11, 12] have also developed predictive models for chip flow and cutting forces. Redetzky [13, 14, 15] has performed considerable work on modeling of the cutting forces and chip flow in machining.

It should be noted that the above predictive models have limited applications in contour turning. Contour machining using a turning tool provides a unique problem due to the continuously changing cutting conditions, effective tool geometry (varying depth of cut, major cutting edge angle and minor cutting edge angle due to varying geometry of contour surface as shown in Figure 1), the resulting wide variations in chip flow, and the consequent chip-form and surface finish issues due to the geometric variations along the length of cut.

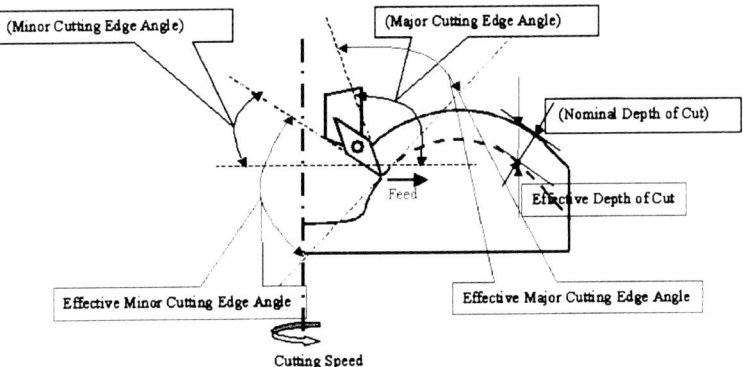

Figure 1: Variation of Effective Depth of Cut, Effective Major Cutting Edge Angle and Effective Minor Cutting Edge Angle Along the Length of Contour Profile

Redetzky's model [16, 17] and the corresponding predictive program were developed for predicting cutting forces and chip flow in straight turning. However, since this model provides a scientific and robust methodology for the prediction of cutting forces and chip flow for varying work material - tool material combinations as well as for a wide range of cutting conditions and tool geometry, it was proposed to extend this model to cover the contour turning case where the effective depth of cut and the effective tool geometry are continuously changing.

1.1 Predictive Model for Cutting Forces and Chip Flow in Contour Turning

Redetzky's predictive model [18] for cutting forces and chip flow is based on the integration of two distinct sub-models:

a) The geometric model which defines the complete geometry of the machining operation based on the cutting conditions (cutting speed, feed and depth of cut) and the tool geometry (cutting edge angle, rake angle, inclination angle and nose radius); and

b) The force model, which establishes the force coefficients for a work material -cutting tool combination as a function of the cutting conditions and tool geometry, based on limited single edge cutting experiments.

These two sub-models are finally integrated to predict the cutting forces for machining operations with a nose radius tool. The calculated cutting forces are also used in predicting the chip side-flow angle based on the effective direction of the resultant friction force on the rake face of the cutting tool.

1.1.1 Redetzky's Geometric Model and Force Model

The fundamental assumption of the geometric model is that the active cutting edge is treated as a series of small single cutting edges. Therefore, the geometric model is based on the division of the cut area A as a whole into regions, which are further subdivided into small elemental cut areas dA (Figure 2). Later, when integrated with the force model, the elemental cut areas develop force elements at each elemental width db and at each elemental area of cut dA of the active cutting edge. The cut areas as well as other geometric parameters are located within the reference plane P_r (Figure 3).

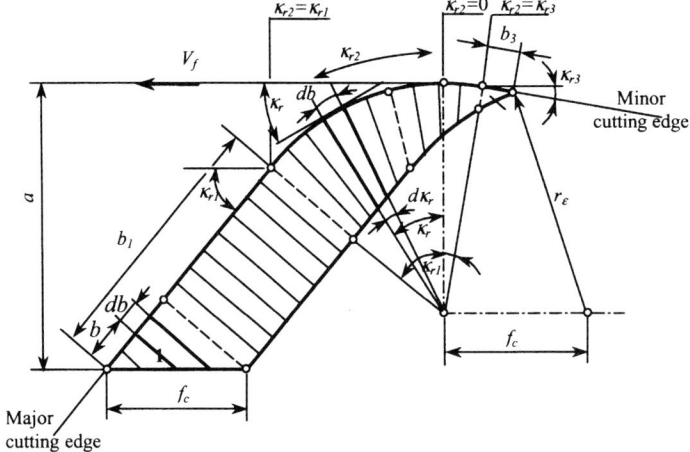

Figure 2: The Undeformed Cut Area and Associated Geometric Parameters [18]

D002/015 © IMechE 2002

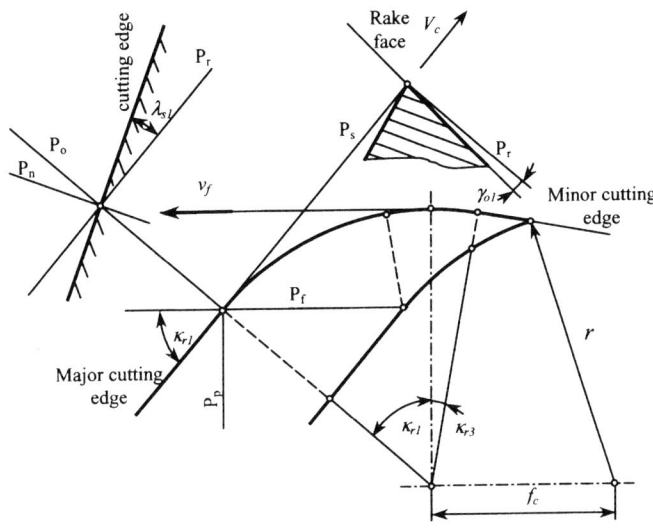

Figure 3: Planes for Measurement of Tool Angles [18]

The fundamental assumption of the force model is that the force behavior in single edge machining (with a width of cut b) can be applied to each elemental width of cut (db) in machining with nose radius tools [18]. Using this model, the secondary "cut area" force elements in the x-y-z directions (F_{Ax}, F_{Ay}, and F_{Az}) by the tool dynamometer can be calculated.

1.1.2 Redetzky's Model for Predicting Chip Flow

Figure 4 shows the basic orientation of the force components and the method for converting them into force components on the rakeface. The basic principle of collinearity of the rakeface forces and the chip flow is used in the prediction of the chip side-flow angle. It is important to note the difference between the chip flow angles η and η_c. The angle η is measured in the reference plane P_r, whereas η_c is measured on the rakeface. The parameters η_c and F are defined within the rake face from their components in planes P_s and P_n. They result from the partial "area of cut" forces F_{Aj} ($j = x, y, z$).

First, the projection of F_{Ax} and F_{Ay} into the directions of planes P_n and P_s results in the force components F_{sr} and F_{nr} as follows:

$$F_{sr} = -F_{Ax} \cos \kappa_{rl} + F_{Ay} \sin \kappa_{rl}$$
$$F_{nr} = F_{Ax} \sin \kappa_{rl} + F_{Ay} \cos \kappa_{rl} \tag{1}$$

where κ_{rl} is the cutting edge angle.
Then, projecting these components and the additional component F_{Az} into the directions of planes P_s and P_n within the rake face, we get (Figure 4):

$$F_s = F_{sr} \cos \lambda_{sl} + F_{Az} \sin \lambda_{sl}$$
$$F_n = F_{nr} \cos \gamma_{nl} + (F_{Az} \cos \lambda_{sl} - F_{sr} \sin \lambda_{sl}) \sin \gamma_{nl} \tag{2}$$

Thus, we get the chip flow force:
$$F = (F_s^2 + F_n^2)^{1/2} \qquad (3)$$
and the resultant chip flow angle on the rakeface
$$\eta_c = \tan^{-1} (F_s/F_n) \qquad (4)$$
We now transform η_c into η (in the reference plane P_r) by using:
$$\eta = \tan^{-1} [(\tan \eta_c \cos \lambda_{s1} - \sin \gamma_{n1} \sin \lambda_{s1})/\cos \lambda_{s1}] \qquad (5)$$
Thus the chip flow angle η is predicted by using predicted cutting force and the tool geometry.

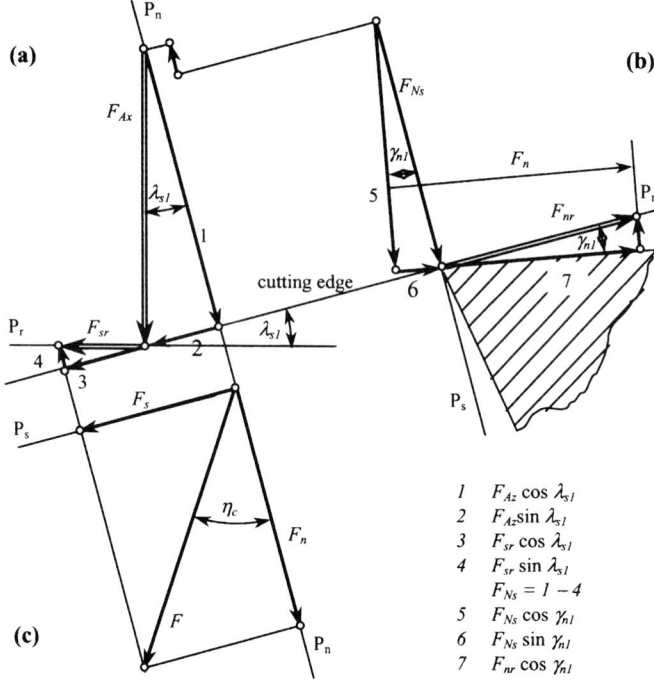

Figure 4: Derivation of the Chip Flow Angle and Chip Flow Force Resulting from the Partial Secondary Forces F_{Ax}, F_{Ay} and F_{Az} (a) Within Plane P_s; (b) Within Plane P_n; (c) On the Rakeface [18]

D002/015 © IMechE 2002

1.1.3 Predicting Chip Flow in Contour Turning

In order to use Redetzky's predictive model for the contour turning case, we need to extend the program to fit the varying effective depth of cut, effective major cutting edge angle and effective minor cutting edge angle due to varying geometry of the contour surface (Figure 1). The cutting parameters needed for chip flow prediction in the extended program include the normal tool rake angle, normal tool inclination angle, cutting speed, tool nose radius, nominal depth of cut, major cutting edge angle, minor cutting edge angle, feed, effective depth of cut, effective major cutting edge angle and effective minor cutting edge angle. The flow chart of the extended program is shown in Figure 5.

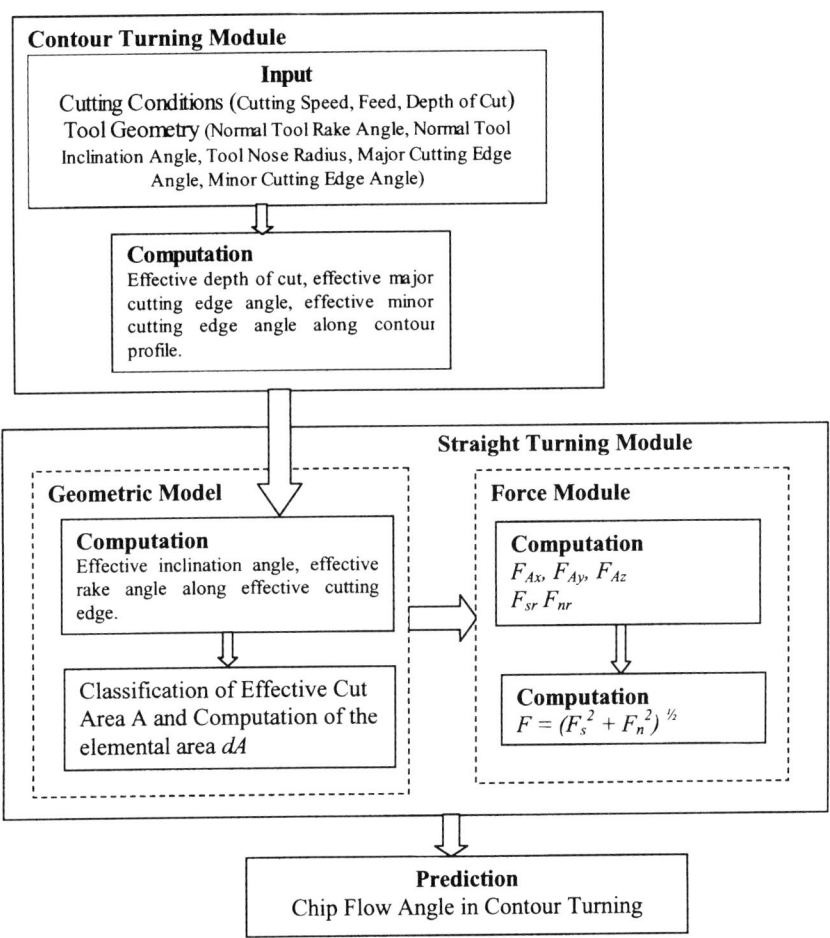

Figure 5: Flow Chart for Predicting Chip Flow in Contour Turning

2. CHIP FORMATION, CHIP CONTROL, AND CHIP FORMER DESIGN FOR DRY CONTOUR MACHINING

A large number of papers in the area of chip formation, chip control and chip former design have recently been published. A historical review of research on chip control was well summarized by Jawahir and van Luttervelt [20]. We can positively identify that the chip former design and tool material play the most important role in chip control for the dry machining case. The tool restricted contact effect is a major influencing factor in chip curl and chip breaking. It is also identified that the present trend in chip former tool developments is toward the design and manufacture of a wide range of specialized chip formers with the specific design features of narrow tool face lands, rake face lumps, curved and wavy cutting edges, curved rake faces and obstruction type back-walls. Each chip former is recommended for a specific machining operation [21].

But in contour turning, due to continuously changing cutting conditions, wide variations in chip flow and the consequent chip-form exist. This poses significant challenges in designing an optimal chip former for a various range of machining conditions.

3. THE APPLICATION: CHIP-FLOW AND CHIP CONTROL IN CONTOUR TURNING OF CAST ALUMINUM WHEELS

3.1 Background
Currently a considerable amount of scrapped wheels are being produced as a result of sub-optimal contour turning processes. Long, snarled, stringy, and uncontrollable hard chips are produced in the form of long ribbons or long spiral coils which cause hair-line scratches on the contour surface of these wheels, particularly along the pocket edges of the wheel surface within the windows regions. This paper is especially motivated by this problem arising in the contour turning of aluminum alloy wheels for the automotive industry.

3.2 Identifying the Problem Region in Contour Machining Aluminum Alloy Wheels
Using the wheel geometry and CNC code, the radial profile of the wheel was constructed (Figure 6 and Figure 7). After a careful study of the resulting contour profiles, both in roughing and finishing operations, the varying effective depth of cut along the roughing and finishing profiles were established (Figure 8 and Figure 9). Also, the associated varying effective major and minor cutting edge angle were calculated (Figure 10 and Figure 11).

Analysis of the effective contour profiles and the corresponding tool geometry variations shown in Figure 6 through Figure 11 can be summarized as follows:

- A continuous cutting process exists in the rim region of the wheel.
- The effective depth of cut and the associated effective cutting edge angle change very rapidly at the rim region of the wheel.
- The chip formation in the rim region is characterized by the production of continuous, long and snarled chips that are difficult to control. The calculation of the chip flow direction along the roughing profile in the rim region, as well as the high speed filming studies (Figure 12), showed that the chip flows toward to the windows of the wheel, thus producing scratches on the wheel surface in the windows region.

D002/015 © IMechE 2002

- The study of the chip flow patterns, involving the use of high speed filming techniques, confirms the chip flow direction towards the windows where the uncontrollable chip builds up and scratches the wheel surface as shown in Figure 13. This study also shows that the worst case happens in the roughing process (Figure 13).

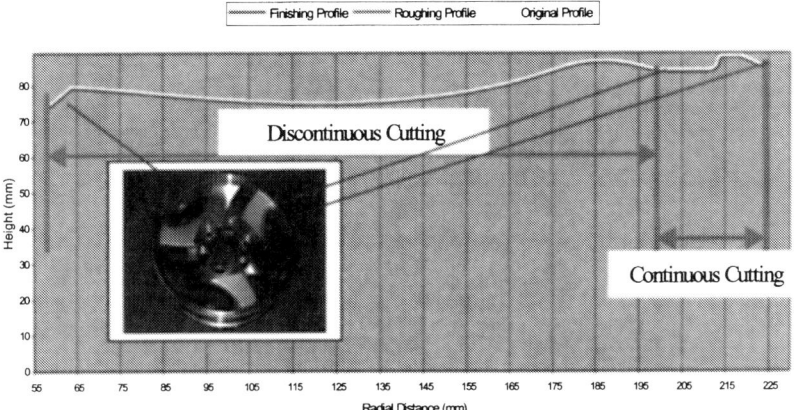

Figure 6: Original, Roughing and Finishing Radial Profiles

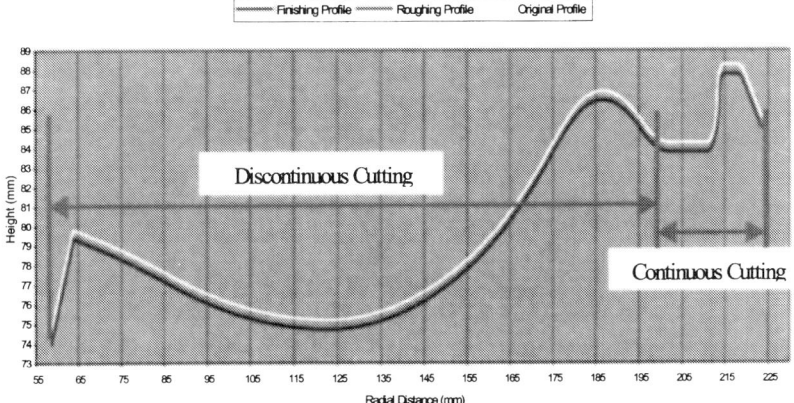

Figure 7: Enlarged View of the Original, Roughing and Finishing Radial Profiles

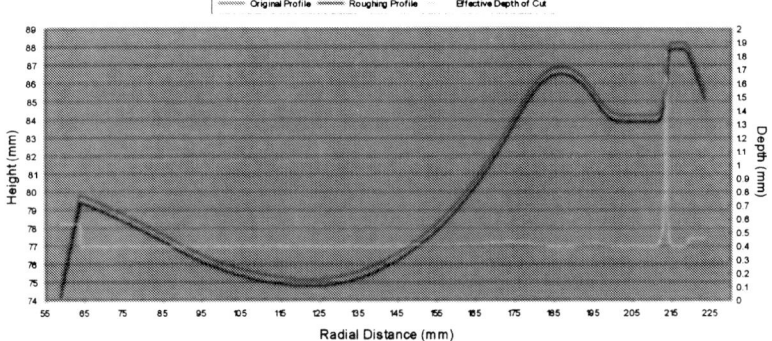

Figure 8: Effective Depth of Cut Along the Roughing Profile

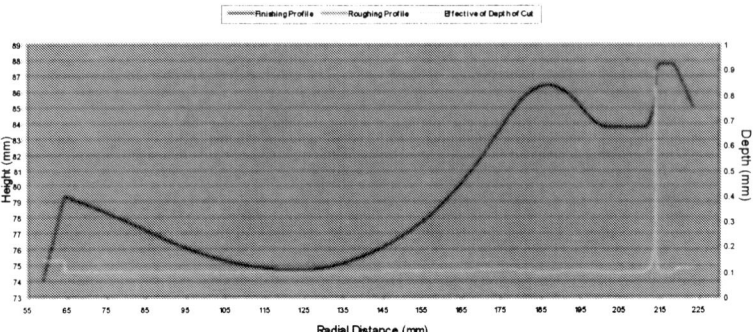

Figure 9: Effective Depth of Cut Along the Finishing Profile

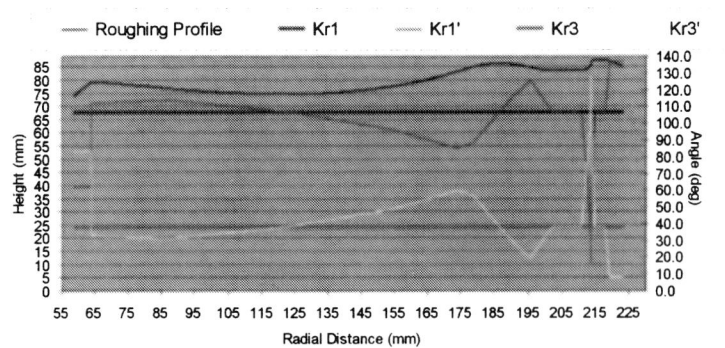

Figure 10: Effective Major and Minor Edge Angles Along the Roughing Profile

D002/015 © IMechE 2002

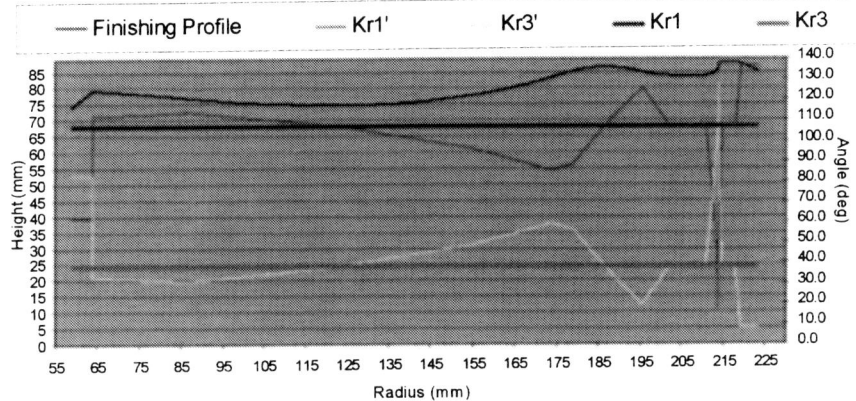

Figure 11: Effective Major and Minor Cutting Edge Angles Along the Finishing Profile

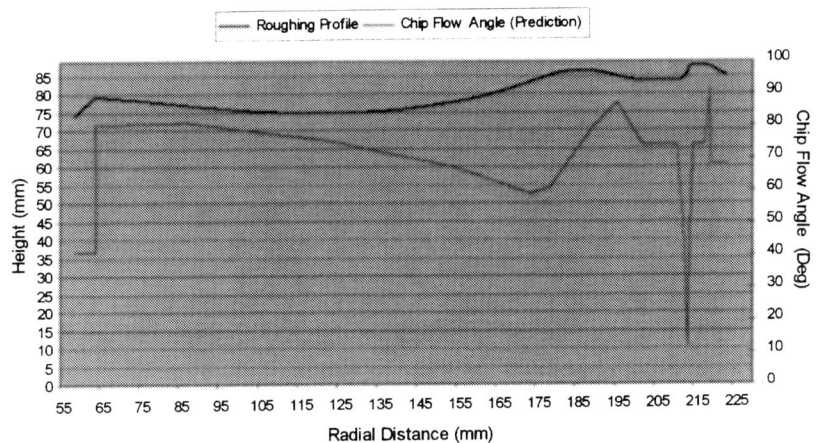

Figure 12: Chip Flow Angle (Prediction) Along the Roughing Profile

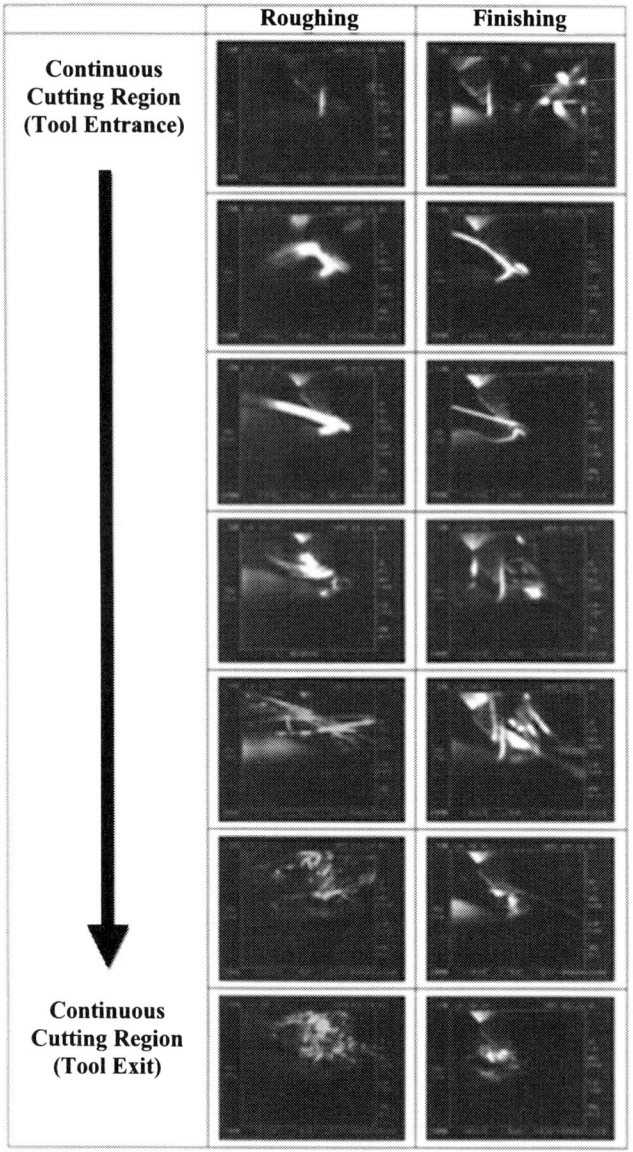

Figure 13: Chip Flow Patterns Along the Profile of the Problem Region in Roughing and Finishing

3.2 Solutions
To avoid the long, continuous and snarled chips scratching the wheel surface, two methods were recommended and were accordingly implemented.

3.2.1 Modification of the Cutting Process
A pre-cut technique was utilized in the sub-region II (Figure 14). The purpose here is to transform the continuous cutting in the problem region to discontinues cutting, thus decreasing the possibility for long chip formation.

Trials on the shop floor show that it does help in decreasing the percentage of scrap wheel production, however, it increases the machining cycle time.

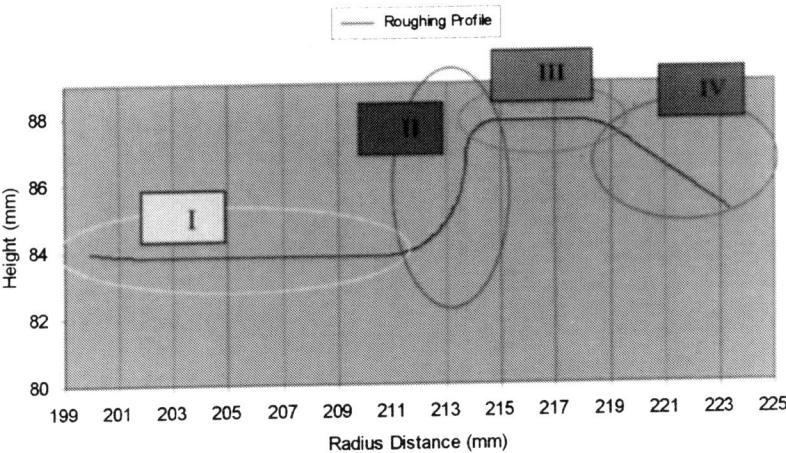

Figure 14: Sub-regions in the Problem Region

3.2.2. New Chip-groove Design
A chip-groove feature was designed and produced using the conventional EDM process on the tool face of the PCD tool insert (Figure 15). The new tool was used in machining to control (i.e., deviate from the windows region of the wheel) the chip flow direction. During the machining process, Region A of the tool insert is used for machining the sub-regions III and I (Figure 14); Region B is used for machining of the sub-region IV (Figure 14); and Region C (Figure 15) corresponds to sub-region II (Figure 14).

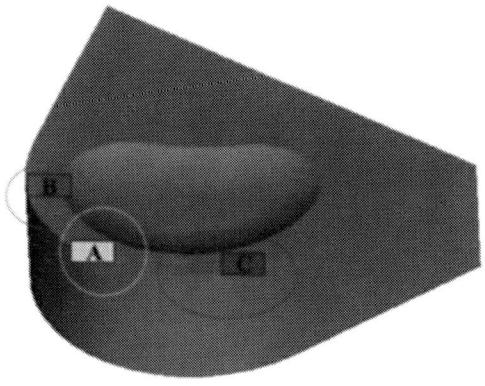

Figure 15: Cutting Regions of the Grooved Tool Insert

A detailed analysis was conducted on the actual machining process using seven newly designed tool inserts. The summarized results are as follows:

(a)   Figure 16 shows the comparison of the chip flow patterns in the problem region in machining with the current flat PCD tool insert and the new grooved tool insert 1 in roughing without pre-cutting. Figure 16 also confirms that the groove tool insert provides a significant improvement in chip control.

(b)   The comparison of the chip flow patterns in the problem region while machining with the grooved tool insert 1 in roughing with and without pre-cutting is shown in Figure 17. In general, no clear difference is noted. In order to make a good judgment on the surface quality of the wheel, further wheel surface observation is needed from quantitative machining trials.

(c)   Figure 18 shows the comparison of chip patterns in the problem region using grooved tool inserts 0, 1, 2, 3, 4, 5, and 6 in roughing with pre-cutting. We encountered some difficulties in obtaining a consistent land length along the active cutting edge due to the EDM process limitations. As a result, some tool inserts came out from the EDM with lower than the required land length. These inserts were expected to pose some problems in machining. As a result, Figure 18 also reveals that a chipping occurs when using the grooved tool inserts 5 and 6. Figure 18 shows that the tip of tool insert 6 was fully chipped off during the machining test. The surface quality of the wheel was also affected in machining with grooved tool inserts 5 and 6.

D002/015 © IMechE 2002

Wheel Profile Sub Region-Tool's Cutting Region	Flat Tool Insert	Grooved Tool Insert 1
I-A		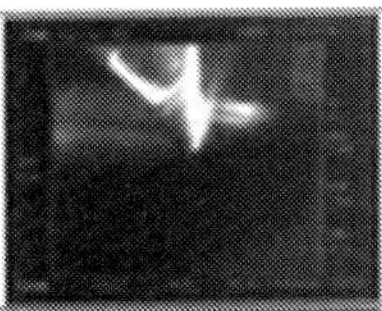
I-A/ II-C	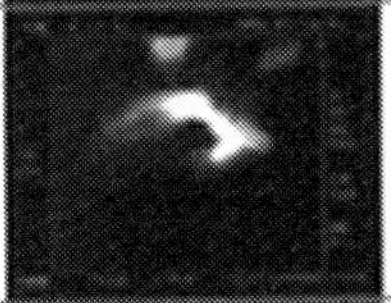	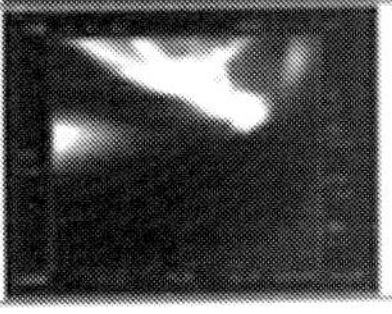
II-C	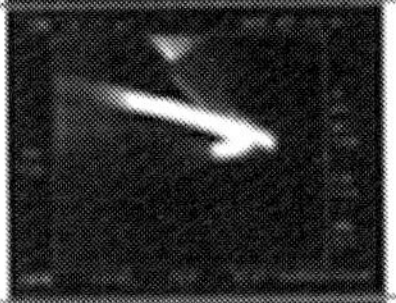	N/A
II-C/ III-A	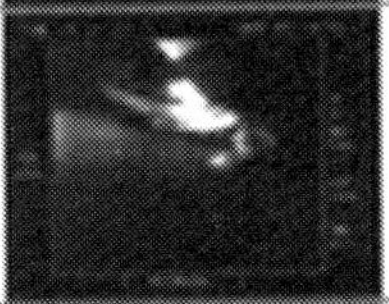	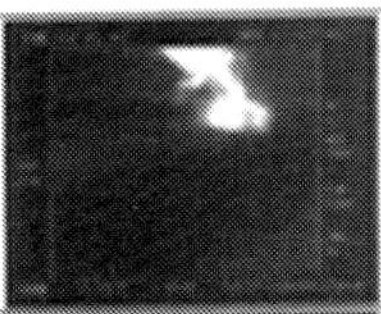
III-A	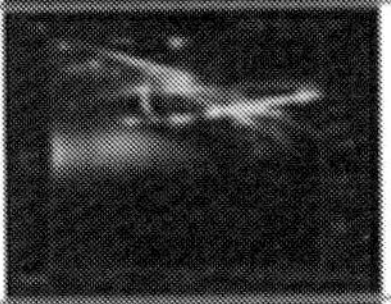	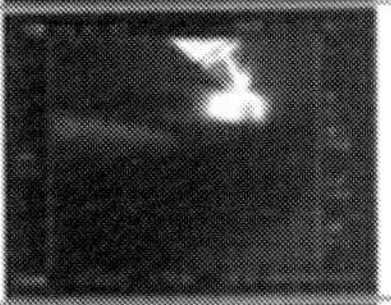
III-A	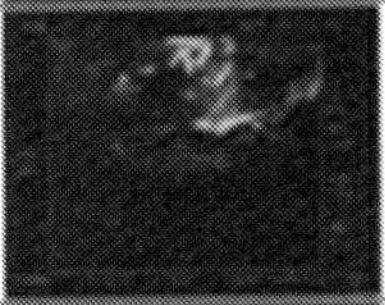	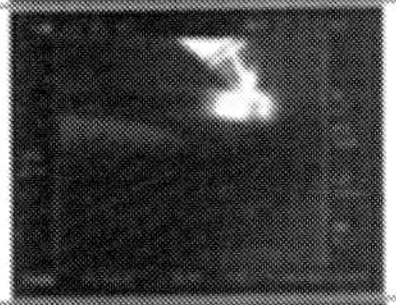
III-A/ IV-B	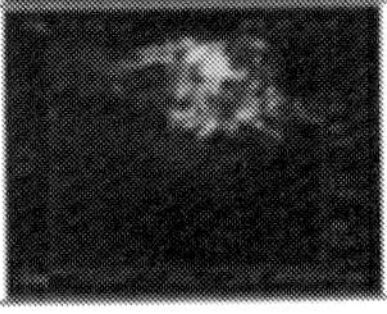	

Figure 16: Comparison of Chip Flow Patterns in the Problem Region by using Flat and Grooved Tool Insert 1 in Rouging Without Pre-Cutting

Wheel Profile Sub Region-Tool's Cutting Region	Grooved Tool Insert 1 (Without Pre-Cutting)	Grooved Tool Insert 1 (With Pre-Cutting)
I-A	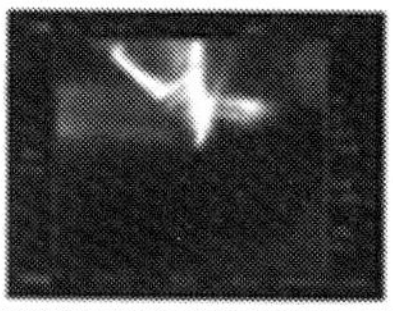	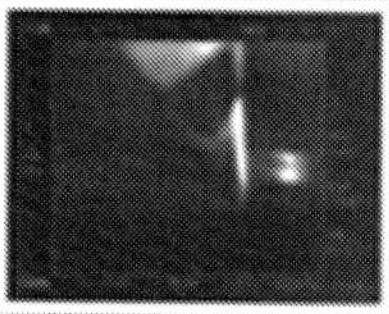
I-A/ II-C	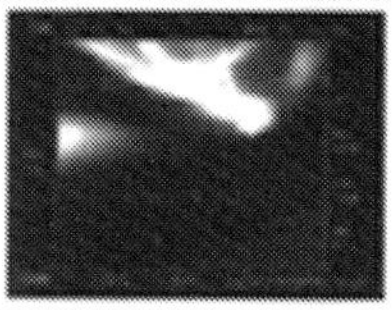	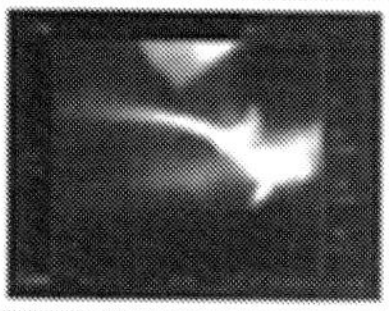
II-C	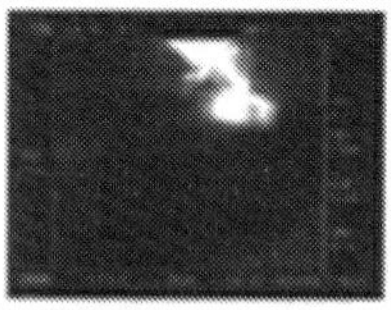	Pre Cutting
II-C/ III-A	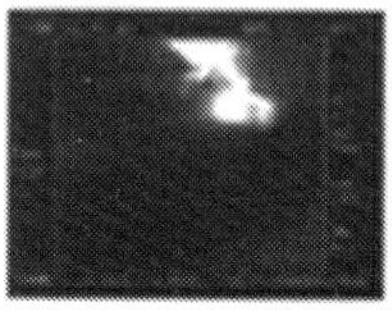	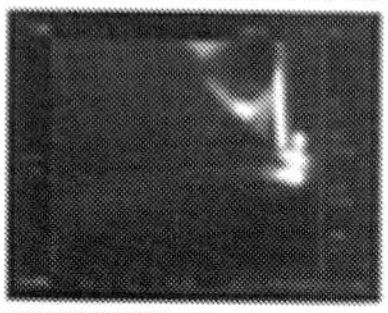
III-A	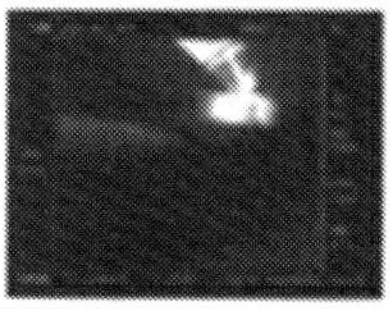	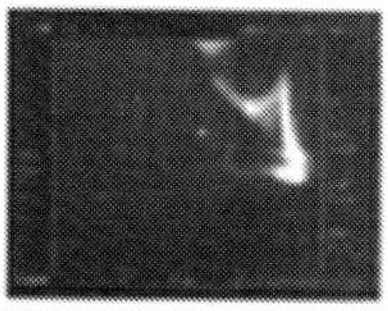
III-A	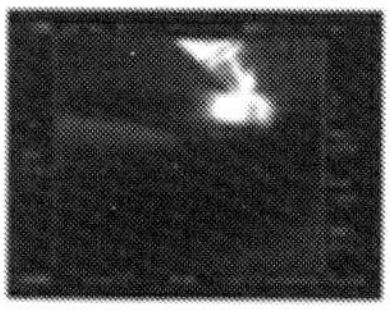	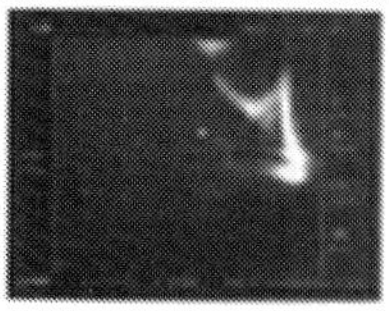
III-A/ IV-B	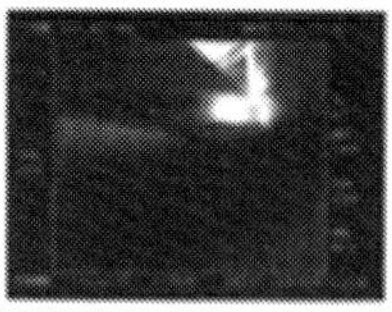	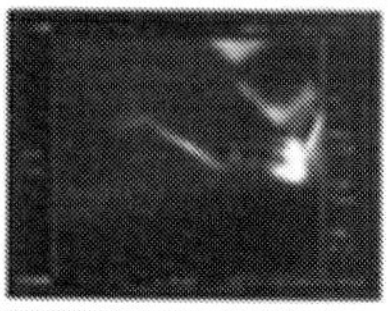

Figure 17: Comparison of the Chip Flow Patterns in the Problem Region by using Grooved Insert 1 in Roughing with Pre-Cutting and without Pre-Cutting

D002/015 © IMechE 2002

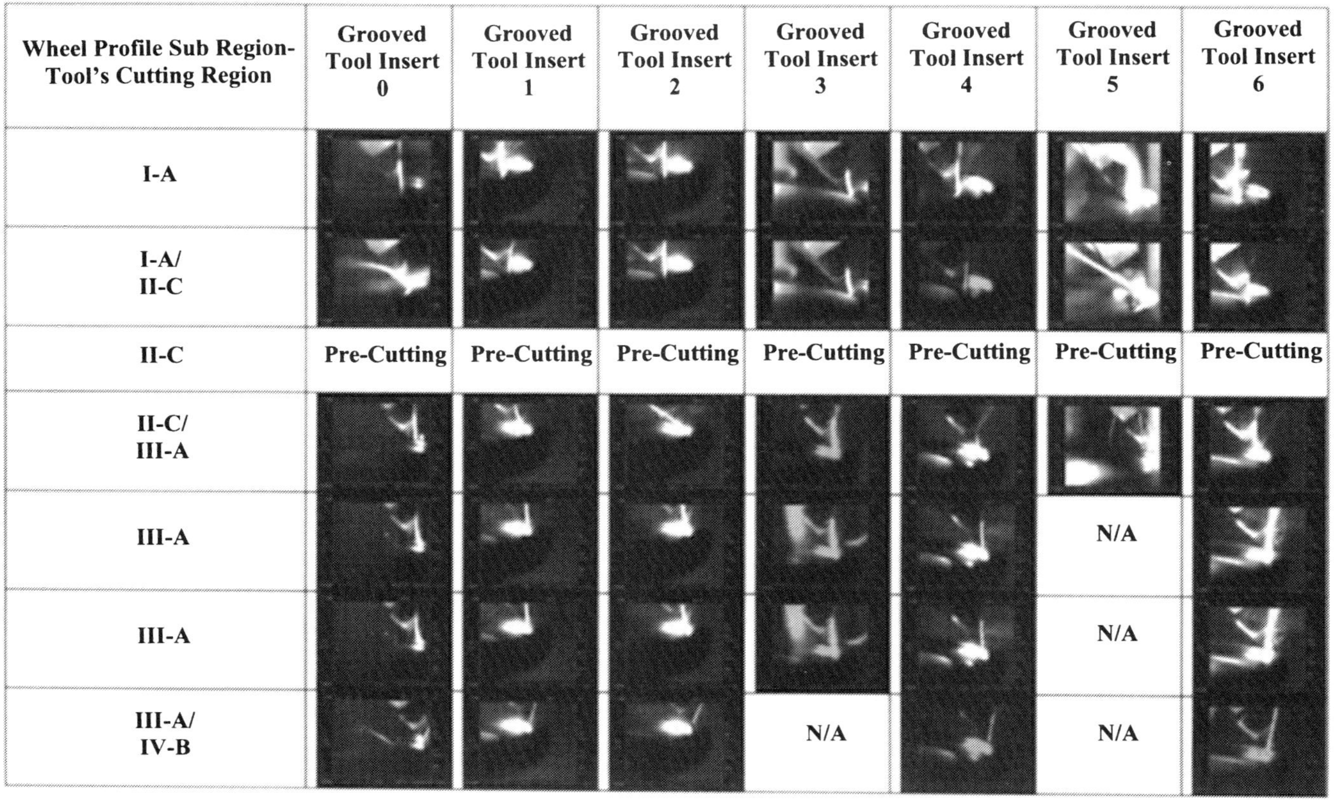

Figure 18: Comparison of Chip Flow Patterns in the Problem Region in Machining with Grooved Tool Inserts 0, 1, 2, 3, 4, 5, and 6 (Roughing with Pre-Cutting)

4. CONCLUSIONS

1. Experimental work has been carried out to investigate the chip flow patterns along the wheel's contour profile in machining. It was shown that at the wheel's outer rim region, labeled as "the problem region", continuous, long and snarled chips are produced. This is due to the rapid changes of the effective depth of cut and the associated cutting edge angle along the tool path in the problem region.

2. A relevant chip-groove geometry for the specific machining operations was developed. The new chip-groove configuration has been produced on PCD tool inserts. The new grooved tool inserts have been tested for a wide range of cutting conditions.

3. Significant improvements in chip flow patterns have been achieved and this helps the inherent chip control problem in contour turning of cast aluminum wheels.

4. It is proposed that new chip-groove configurations can be produced on uncoated carbide tool inserts followed by a suitable diamond coating, resulting in significantly reduced tool insert cost.

5. The anticipated broader functionality of the new tool design is also to offer a coolant-free dry machining process.

REFERENCES

1. **Miller, W. S., Zhuang, L., Bottema, J., Wittebrood, A. J., De Smet, P., Haszler, A., Vieregge, A.** (2000) Recent Development in Aluminum Alloys for the Automotive Industry. *Materials Science and Engineering*, A280, pp. 37-49.

2. **Cole, G. S. and Sherman, A. M.** (1995) Lightweight Materials for Automotive Applictions. *Materials Characterization*, Vol. 35, pp.3-9.

3. **Colwell, L. V.,** (1954) Predicting the Angle of Chip Flow for Single-Point Cutting Tools. ASME, Vol. 76, pp. 199-204.

4. **Stabler, G. V.,** (1951) The Fundamental Geometry of Cutting Tools. *Proc. Instn. Mech. Engrs.*, Vol. 165, pp. 14-21.

5. **Stabler, G.V.,** (1964) The Chip-flow Law and Its Consequences. *Proc. 5th Int. Mach. Tool Des. Res. Conf.*, Pergamon, Oxford, pp. 243-251.

6. **Brown, R. H., Armarego, E. J. A.,** (1964) Oblique Machining with a Single Cutting Edge. *Int. J. Mach. Tool Des. Res.*, Vol. 4, pp. 9-25.

7. **Armarego, E.J.A.,** (1983) General Mechanics of Cutting Analyses for Machining with Plane Faced Cutting Tools. *1st IMCC*, South China Inst. of Tech., Guangzhou, China, pp. 11-35.

8. **Armarego, E. J. A., Samaranayake, P.,** (1997) Predictive Performance Modelling of Turning Operations for Modern Computer Based Manufacturing. *CIRP International Symposium*, Hong Kong, pp. 304-312.

9. **Lin, C. G. I., Mathew, P., Oxley, P. L. B., Watson, A. R.,** (1982) Predicting Cutting Forces for Oblique Machining Conditions. *Proc. Instn. Mech. Engrs.*, Vol. 196, pp. 141-148.

10. **Young, H.T., Mathew, P., Oxley, P.L.B.,** (1987) Allowing for Nose Radius Effects Predicting the Chip-Flow Direction and Cutting Forces in Bar Turning. *Proc. Inst. Mech. Engrs.*, C201(C3), pp. 213-216.

11. **Arsecularatne, J. A., Mathew, P., Oxley, P. L. B.,** (1995) Prediction of the Chip Flow Direction and Cutting Forces in Oblique Machining with Nose Radius Tools. *Proc. Instn. Mech. Engrs.*, Vol. 209, pp. 305–315.

12. **Arsecularatne, J. A., Fowle, R. F., Mathew, P., Oxley, P. L. B.,** (1996) Prediction of Cutting Forces and Built-up Edge Formation Conditions in Machining with Oblique Nose Radius Tools. *Proc. Instn. Mech. Engrs.*, Vol. 210, pp. 457–469.

13. **Redetzky, M.,** (1979) Berechnung der Vorschubkraft und der Passivkraft und ihre Nutzung in technischen Restriktionen zur Sicherung der Fertigungsgenauigkeit (Prediction of feed force and passive force and their application in technical constraints for ensuring precision manufacturing). Dresden Techn. Univ., Dr.-Ing. Diss.

14. **Redetzky, M.,** (1985) Eine neuartige Methode zur einheitlichen Berechnung aller Komponenten der Spanungskraft (A novel methodology for an universal prediction of all components of the resultant cutting force). *Wiss. Z. Techn. Hochsch.* Karl-Marx-Stadt, Vol. 27(6), pp. 951-963.

15. **Redetzky, M.,** (1987) Bestimmung der Spanablaufrichtung beim Drehen mit Hartmetall (Prediction of the chip-flow direction in turning with carbide tools), *Wiss. Z. Techn. Univ.* Magdeburg, Vol. 31(7), pp. 40 – 45.

16. **Redetzky, M., Balaji, A.K. and Jawahir, I. S.,** (1998) Predictive Modeling of Cutting Forces and Chip Flow in Machining Operations. *CRMS Research Report*, University of Kentucky, 1998.

17. **Redetzky, M.,** Programm KOSA. Einheitliche Berechnung aller Komponenten der Spanungskraft und der Spanablaufrichtung beim Realschnitt (Universal prediction of all components of the resultant cutting force and of the chip-flow direction in oblique machining with nose radius tools). 1985/1995/1998.

18. **Redetzky, M., Balaji A. K., and Jawahir, I. S.,** (1999) Predictive Modeling of Cutting Forces and Chip Flow in Machining with Nose Radius Tools. *Proc. 2nd CIRP Int. Workshop on Modeling of Machining Operations*, Nantes, France, pp. 160-180.

19. **Balaji A. K., and Jawahir, I. S.** (2001) A Machining Performance Study in Dry Contour Turning of Aluminum Alloys with Flat-Faced and Grooved Diamond Tools. *Machining Science and Technology*, Vol. 5(2), pp. 269-289.

20. **Jawahir, I. S. and van Luttervelt, C. A.,** (1993) Recent Developments in Chip Research and Applications, *Annals of the CIRP*, Vol. 42 (2), pp. 659-693.

21. **Jawahir, I. S.,** (1988) The Tool Restricted Contact Effects as Major Influencing Factor in Chip Breaking: An Experimental Analysis. *Annals of the CIRP*, Vol. 37 (1), pp. 121-126.

D002/015 © IMechE 2002

Advances in the
Electronics Sector

Development of a generic model for life-cycle inventory (LCI) of upstream processes in life-cycle assessment (LCA) of electronic products

A S G ANDRÆ and **J LIU**
Division of Electronics Production, Chalmers University of Technology, Gothenburg, Sweden

Abstract

Considerations of cost, high speed, high reliability, high manufacturability and environmental compatibility have to be taken into account in future product development of electronic products. To simulate the environmental compatibility, for example a Life Cycle Assessment (LCA) can be performed. An electronic product usually consists of hundreds to thousands of electronic and mechanical components. To perform an LCA is very time consuming if a Life Cycle Inventory (LCI) is done for every single component. This is not needed because the unit processes in the upstream product system in some cases are the same for the ingoing components. In this paper a generic model for LCI is developed and applied to one product system (a business telephone). In this model, the components are divided into main groups and then into sub-groups that result in process modules for unit processes that are similar for the ingoing components. There already exist some models that describe the LCI of electronic products in other ways. This model takes the LCI strategy for the manufacturing phase one step further by identifying which processes constitute part of the common denominator of the upstream product system and proposes which ones should be inventoried first. The developed model makes it possible to obtain a higher resolution and level of understanding compared to earlier models with respect to both components and processes.

1 INTRODUCTION

When the environmental problems relating to the manufacture and use of products were first addressed, attention was given to specific substances and specific impact media but the problems are caused by emissions from many sources. This has led to the development of product oriented environmental policies, such as the Integrated Product Policy (IPP) for sustainable development by the European Commission. The goal of such policies is the ecological optimisation of products and services by implementing different instruments from LCA or design for environment to eco-labelling or product declarations and to refurbishing or recycling of discarded products (1).

The European Commission has also published a working paper for a directive addressing the impact on the environment of Electrical and Electronic Equipment (EEE) (2).

The proposal aims to protect the environment by specifying "ecological-design" requirements for all electrical and electronic equipment. If adopted, it will require that manufacturers, in the product design process, assess and take into account environmental attributes at each stage of a product's life cycle in order to be able to market the product within the EU (3).

The awareness about the environmental effects caused by products and also the interests in the development of methods to better understand and reduce these effects have increased. Life Cycle Assessment (LCA) is one of the methods under development for this purpose. LCA is a model for assessing the environmental aspects and potential effects associated with a product, by

- Compiling an inventory of relevant inputs and outputs of a product system;
- Evaluating the potential environmental impacts associated with those inputs and outputs;
- Interpreting the results of the inventory analysis and impact assessment phases in relation to the objectives of the study.

In a LCA study, the potential environmental effects throughout a product's life (i.e. cradle-to grave) from raw material acquisition through production, use and disposal is investigated. The general categories of environmental effects needing consideration include resource use, human health, and ecological consequences (4). LCA is an analytical tool based on physical quantified values which support decision-making.

The life cycle inventory is the most time consuming part of a LCA study, especially for electronic products as the product systems are complex. The amount of data needed to reach a high resolution in a LCA study is very large, but in principal it would be possible to collect the data for every unit process (5). (Unit process is defined in section 2). Economical restraints, lack of time and the number of organisations involved in the data gathering would in effect however stop the attempt.

Concern has arisen that the upstream processes for electronic products are more important than could be expected from the LCA studies. Moreover the yield affects the results for intermediate parts and components, i.e. for the conversion of silicon disks (wafers) into silicon microchips, for the production of wafers from crystalline silicon columns and for the production of thin-film-transistor liquid-crystal displays (5), (6).

Spielmann and Schischke points out that LCA for the production of electronics should focus on evaluation of generic data for suppliers of components, a modularisation of the infrastructure processes in semiconductor manufacturing, frequent up to date evaluation of the semiconductor processes and knowledge exchange between experts on how to make the LCA methodology leaner (7).

This paper describes a model for structuring and categorization of the data needed for LCA of electronic products built using conventional packaging technology in more depth than earlier

D002/014 © IMechE 2002

studies. In particular the grouping of components has been refined compared to the classification in (8).

The component types are divided into main groups and subgroups and then the number of modules needed to make a more accurate assessment of the environmental load of the upstream manufacturing processes is estimated. The model makes it easier to know for which unit processes LCI data have to be gathered.

2 DEFINITIONS OF A PRODUCT SYSTEM AND A UNIT PROCESS

A product system is a collection of unit processes connected by flows of intermediate products. Elementary flows and product flows and intermediate products flow across the product system boundaries. An example of a product system for an integrated circuit (IC) can be seen in figure 1. A unit process is a process where chemical and/or physical changes occur. An example of a set of unit processes can be seen in figure 2. (9). A unit process is linked to a product system by a product flow, to another unit process by an intermediate flow and to the environment by an elementary flow.

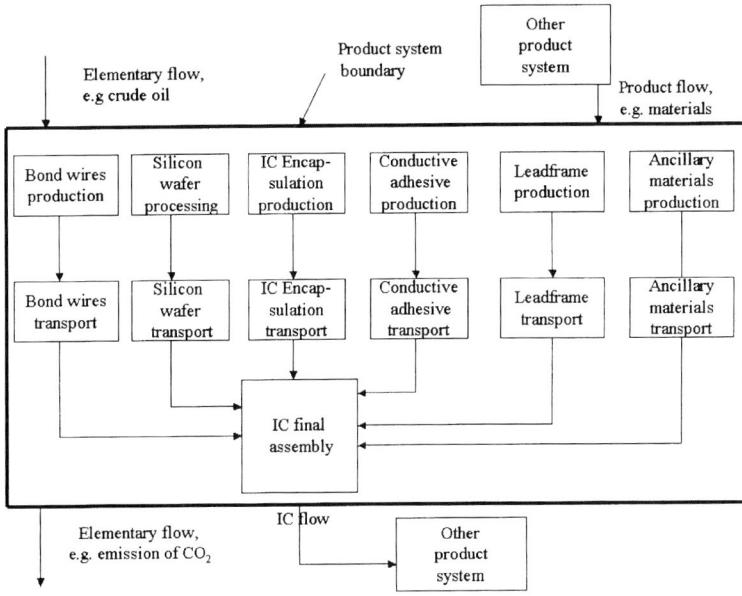

Figure 1. Example of a product system.

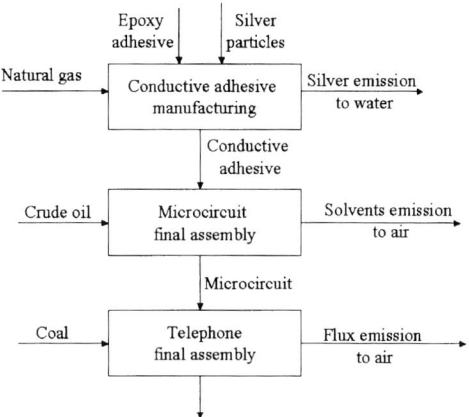

Figure 2. Example of a set of unit processes

3 EARLIER STUDIES AND APPROACHES

Herrmann et al (10) have performed a LCA study using a generic modules approach in order to categorise components of a product system. In their model, the categorization and modules division is rather coarse without knowledge of the detailed component and processes. Therefore, their LCA may be unreliable.

The conclusion is that the result gives a relatively detailed impression of the environmental effects of the different life cycle phases and the different components of the electronic product.

The Canadian National Office of Pollution Prevention, 2000, (11), have together with Nortel performed a LCA of a business telephone where ready-made modules for environmental impact for a complete module, e.g. resistors, from Ecobalance's database on electronic parts are used. Complex components such as microcircuits and printed circuit boards were modelled by collecting data from suppliers.

The conclusion is that the most challenging aspect of a LCA is the access to representative, reliable and accurate data. The data are not collected centrally or adapted easily for LCA purposes. Especially, data collection from primary sources can be time consuming.

Pollock et al (12), have performed a LCA of an inkjet print cartridge where seven suppliers were asked for data and the in-house processes were modelled in detail. The conclusion is that the study was completed within the range of the product development cycle though the data collection was extensive. The database compiled allows full LCA of similar products to be done in much shorter time.

Satake et al (13), made a LCA of CD-ROM drives where data for electronic parts were gathered from a government publication. The energy consumption of electronic parts production, as a function of cost of electronic parts, was shown to be almost linear. The conclusion was that the production processes had the largest contribution to the total environmental load.

Scheller et al (14), has made an LCA of a pager and for all the important processes in the preproduction it was possible to contact the main suppliers and seek data. The conclusion was that manufacturing contributed significantly to the overall environmental impact. A lack of information about the production of electronic components was however found.

Van der Wel et al (15), made process inventories and an environmental evaluation of the manufacturing of ceramic multilayer capacitors, surface mounted device (SMD) and conventional resistors. The main input materials and ancillaries were identified, but LCI data for the supplied materials were unavailable, and studies of a public database and the literature were used to fill the data gaps.

Ueno et al (16), provide an original calculation of the LCI data for each electronic components industry. The data show the amount of input energy and emissions per yen of production yield. It was concluded that it is possible to calculate the LCI data for any electronic component by multiplying the LCI data for the industry by the price of the component. The use of input-output tables representing Japan's economic sectors was used to obtain the LCI data. A disadvantage with this method is the difficulty in analysing which stage has the most influence on the environment.

Taiarol et al (17), has made a "gate-to-gate" LCI of a microcircuit, where one conclusion is that only a few data on materials and processes are available for the high-tech materials and processes in the electronic industrial sector. The system studied was subdivided into elementary production steps that were transformed into LCA modules. LCA modules allow the study of IC devices whose production process parameters differ. Material production energy density (gross energy, MJ/kg) is introduced as an indicator of how high the technology level of the production process is.

(8), (11)-(14) and (16) present LCA case studies in which LCI data for electronic components have been collected from suppliers. However, no generic models for LCI data collection have been established. None of the above authors have tried to investigate how large the data collection need really is. Cleary, the biggest issue in the LCA of electronic production is unavailability of relevant data. An overall impression would make it easier to see for which processes data had to be collected. The LCI modules could be made more accurate and co-operation between companies producing the same components, e.g. passive cold be enhanced.

4 MODELLING

In order to address the issues identified in section 3 a model was made to see how large the number of possible unit processes is and which of them are alike.
The present model divides the electronic product into different component groups based on function. An example of the model structure can be seen in figure 3.

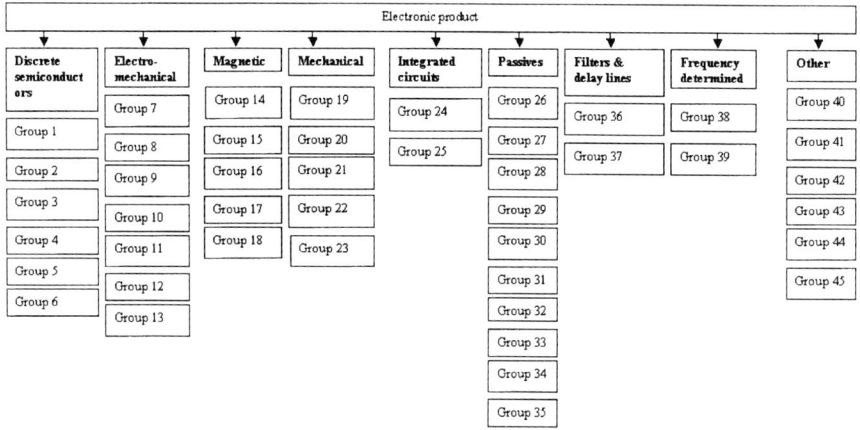

Figure 3. Example of a model for LCI data collection where the concept of using main groups and sub-groups of product systems for electronic components is used.

In this model structure, the LCI data collection is divided into nine main groups and 45 sub-groups. The subgroups are: 1. Display units, 2. Diodes 3. Indicators, 4. Opto couplers, 5. Thyristors, 6. Transistors, 7. Component holders, 8. Connectors, 9. Fuses, 10. Relays, 11. Switches, 12. Microphones, 13. Loudspeakers, 14. Inductors, 15. Chokes, 16. Thread spirals, 17. Transformers, 18. Coil formers, 19. Plastics mechanics, 20. Metallic mechanics, 21. Composite mechanics, 22. Printed circuit boards, 23. Cables, 24. Standard IC's, 25. ASIC's, 26. Metallised paper capacitors, 27. Metallised plastics capacitors, 28. Ceramic capacitors, 29. Electrolytic capacitors, 30. Large capacitors, 31. Potentiometers, 32, Thread wounded resistors, 33. Resistor networks, 34. Varistors, 35. Thermistors, 36. Ferrite rods, 37. Delay lines, 38. Quartz crystal units, 39. Oscillators, 40. Adaption units, 41. Flexible disks, 42. Lamp panels, 43. Batteries, 44. Buzzers, 45. Antennas.

These 45 subgroups can be if necessary divided into their sub parts and then more for inventorable unit processes are identified. Examples of these unit processes are wafer processing and leadframe manufacturing (etched or stamped leadframe manufacturing) which are needed to produce diodes, transistors, and microcircuits. Unit processes needed to manufacture raw materials are outside the scope of this study. Wafer manufacturing can be divided into a number of unit operations but for this model one factory site is defined as one unit process. The leadframe manufacturing may include a plating process but then the leadframe factory is regarded as one unit process.

The LCI model for data collection described in this way is capable of making a detailed and quantifiable LCA.

5 ILLUSTRATIVE EXAMPLE

Here an example of using the modelling approach developed is applied to a Dialog 3213 Digital System Telephone from Ericsson Enterprise AB.

D002/014 © IMechE 2002

Figure 4. Digital System Telephone – Dialog 3213

The telephone Dialog 3213 consists of 247 physical components which can be divided into the following main groups, as can be seen in table 1: Discrete semiconductors, Electromechanical, Magnetic, Mechanical, Integrated circuits, Passives, Filters and delay lines and frequency determined components. The main groups are formed based on the function that the components perform. The names of sub-groups in bold type in table 1 are those found in the Dialog 3213.

Table 1. Dialog 3213 divided into main and sub-component groups

Main group	Sub-groups	No. of components in Dialog 3213
Discrete semiconductors	**Diodes, Display units**, Indicators, Opto couplers, Thyristors, **Transistors**	47
Electromechanical	**Component holders, Connectors,** Fuses, Relays, **Switches, Microphones, Loudspeakers**	9
Magnetic	**Inductors**, Chokes, Thread spirals, **Transformers**, Coil formers	6
Mechanical	**Plastic mechanics, Metallic mechanics, Printed circuit boards, Cables,** Composite mechanics	54
Integrated circuits	**Standard IC's,** ASIC's	10
Passives	Metallised paper capacitors, **Metallised plastics capacitors, Ceramic capacitors**, Electrolytic capacitors, Large capacitors, Potentiometers, **Thread wounded resistors**, Resistor networks, Varistors, Thermistors	120
Filters and delay lines	Ferrite rods, Delay lines	0
Frequency determined components	**Quartz crystal units**, Oscillators	1
Other	Adaption units, Flexible disks, Lamp panels, Batteries, Buzzers, Antennas	0

The components in the sub-groups column seen in table 1 can in turn be divided into sub-part groups to start the decision making process as to the number of unit processes needed to describe the upstream product system.

Table 2. Unit processes needed to produce the sub parts of a microcircuit

Component	Sub part	Unit processes in the upstream
Standard IC	Chip	Wafer processing, Raw wafer manufacturing, monocrystalline silicon manufacturing
	Encapsulation	Encapsulation manufacturing
	Leadframe	Stamped leadframe manufacturing, Etched leadframe manufacturing
	Bond wires	Bond wire manufacturing
	Conductive Adhesive	Conductive Adhesive manufacturing

Table 2 shows one example of the modelling approach adopted and the same procedure has been applied to the rest of the components included in Dialog 3213 the resulting of a total of 121 factories to be considered. The number of unit processes generated by component assemblies, e.g. the final assembly of the display unit is included, which make 19 from the groups in table 1 and 102 from processes further upstream. For Standard IC types these unit processes for factories are defined: Final assembly of Standard IC, Processing of the raw wafer, Manufacturing of the raw wafer, Manufacturing of monocrystalline silicon, Etched leadframe manufacturing, Stamped Leadframe manufacturing, Bond wire manufacturing, Conductive adhesive manufacturing and Encapsulation manufacturing. For this study no LCI data has been captured but it is meant that a user of the model should collect data from each factory.

6 DISCUSSION

Raw wafer manufacturing and monocrystalline silicon manufacturing are examples of unit processes that are the same for diodes, transistors and integrated circuits and all other components that use silicon chips. In figure 5 the similarities for diodes, transistors and integrated circuits is shown and in figure 6 an example of a network diagram for which unit processes that are similar for diodes, transistors and integrated circuits can be seen. Dotted boxes in figures 6 indicate the exclusion of unit processes from the study. The wafer processing unit processes are unique but the difference between factories has to be investigated. The size of the wafer processed and the number of process steps are a factors that may lead to more than one LCI module. For display units, whose sub-parts also to some extent include microchips, the raw wafer and monocrystalline silicon manufacturing can be used. The manufacturing of metallurgical silicon from sand is not included here as it regarded as a raw material.

An LCI module for manufacturing of encapsulation materials for microcircuits, diodes and transistors and similar components could be standardised. Conductive adhesive is expected to be an important ingredient of future electronics and the manufacturing of the adhesive used to attach the chip inside the circuits is one unique unit process (18).

D002/014 © IMechE 2002

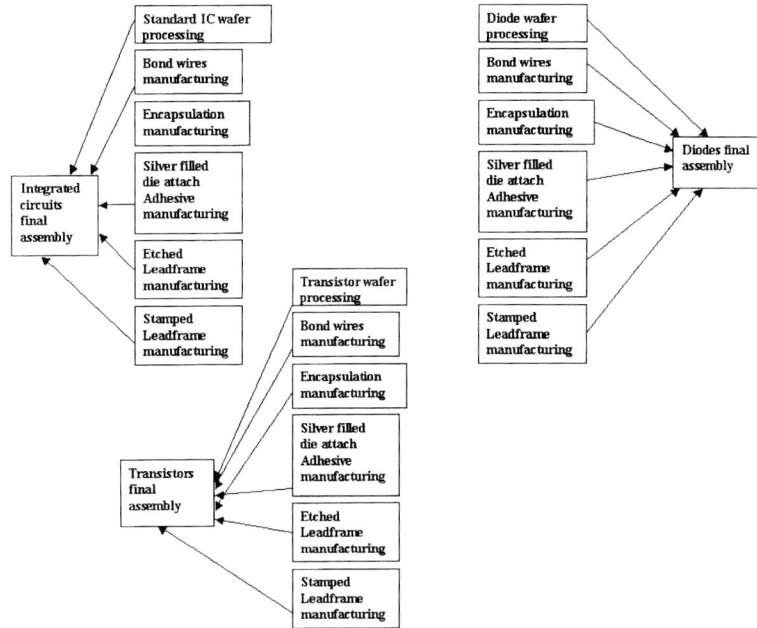

Figure 5. Flow charts of unit processes occurring in production of three types of silicon containing components

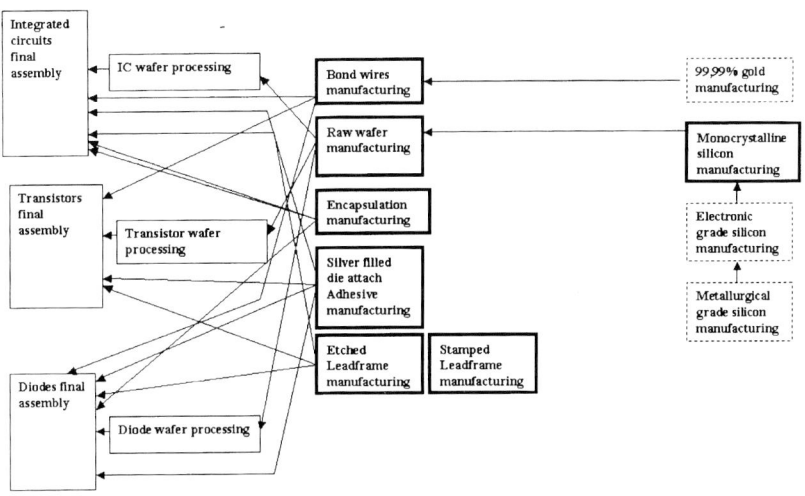

Figure 6. A network diagram of similar unit processes in the upstream processes of integrated circuits, transistors and diodes.

The difference between the thinking of this article and (8)- (17), is that it tries to identify the processes for which data could be collected to support environmental assessment of any electronic product.

The difference compared to other models is that this model tries to identify which unit processes in the upstream processes that are similar and that other "models" make LCI modules for whole components from cradle-to-gate.

The result only reflects one electronic product that moreover does not "use" all of the 45 component groups. More unit processes would be identified if more component groups than the 45 would be identified. Neither does the paper analyse all of 45 component groups.

The resolution of the product system flow chart, which shows which unit processes should be inventoried, has to be higher if the real environmental load of complex products is to be revealed. Reasonable conclusions can be drawn without this model, but it enables the user to have an overview of which process data have to be collected. Small, inexpensive consumer products and expensive, highly complex, information technology systems undoubtly to some extent share the same unit processes in the upstream of their respective product system.

If manufacturers must analyse the environmental impact of each component within their products in order to be able to provide the environmental impact of their product at each stage of its life cycle, this work could help. If this model should be adopted, burdensome financial and administrative costs could be reduced and delays in new product developments would be less likely to occur.

The LCA practitioner thinks in terms of unit processes which have to be "filled" with data to give the environmental load and subsequently the potential environmental effects of the components from which the electronic product is made. Earlier LCA studies performed for electronic products indicate that the wafer processing is the most environmentally loading unit process in the whole upstream, but this must be quantified (17). It is the task of the component manufacturers to coordinate the LCI data work on demand from the producers of electronics. One of the users of the model is the electronics industry that needs to examine its environmental impact.

It is a good way of working to make LCI modules for the electronic components but then the modules must include all relevant unit processes. The general LCA user is interested in the main manufacturing steps of a product or a component and not always all technological steps.

7 CONCLUSIONS

The work presented here can be seen as a development of a sector specific standardisation for the specific product group of electronic products. A development of the ISO standards is needed to eliminate some of the arbitrariness as the methods for doing the LCA of complex products differs from others which consists of fewer materials and components. Today the uncertainties of the results are caused by the choice of the most relevant data, choice of functional unit, cut-off criteria, geographical and technological assumptions, allocation procedures and impacts assessment methods and data (19).

D002/014 © IMechE 2002

LCA studies of products are partly about gathering data for unit processes in the upstream processes. In this article some of the unit processes that are common for a business telephone have been identified. The aim with the work is that the transparency of the product systems for electronic products is increased and that the LCI modules for electronic components will have better accuracy.

It is also important that all ingoing components have their own unique material content declaration, but it would be very time consuming if every component should have its unique LCI data. By identifying underlying unit processes, which are the same for the components, the work can be simplified.

The data gathering need is vast but not as vast as it would be if every component had to be inventoried from cradle-to-gate. The data collection can be minimized if the proposed model is used.

8 FUTURE WORK

The next step would be to identify more unit processes and also look at which are the most common production chemicals in the electronics industry, so LCI modules can be made for them. As an example, about 45 different lacquers, each consisting of 5-10 materials are used for the production of conventional resistors (15).

It should be discussed within the electronics industry how the data gathered and other findings could be shared once the unit processes have been identified.

ACKNOWLEDGEMENTS

The publication of this work has been authorized by Ericsson Enterprise AB. Financial support for this work has been provided by EPROPER (a Swedish National research school for graduate students in the area of electronics production and packaging). Credits go to David Whalley, Helge Kristiansen, Tomas Rydberg, Raul Carlson and Ann-Marie Tillman for useful comments.

REFERENCES

1   http://europa.eu.int/comm/environment/ipp/home.htm

2   http://europa.eu.int/comm/environment/docum/00347_en.htm

3   http://www.sun.com/aboutsun/policy/eco-standard.html

4   ISO (1997) ISO 14040: Environmental management—life cycle assessment—principles and framework.

5   **Rydh C.J.**,(2001) Environmental assessment of battery systems in life cycle management In *Licentiate thesis, Chalmers University of Technology, Gothenburg* pp. 21, 41

6   **Doka G.** (2001) Yield losses in electronics production are significant to LCA. In *Environmental Impact of Telecommunication System and Services 13 th Discussion Forum on Life Cycle Assessment; Lausanne, Switzerland*, pp 9-12.

7   **Spielmann, M., Schischke, K.** (2001) Environmental Assessment in Production of Electronic Components - Possibilities and Obstacles of LCA Methodology. In *Environmental Impact of Telecommunication System and Services 13 th Discussion Forum on Life Cycle Assessment; Lausanne, Switzerland*, pp. 9-21.

8   **Andrae, A. S.G., Östermark, U. and Liu, J.** (2001) Life cycle assessment of a telecommunications exchange. In *Journal of Electronics Manufacturing.* Vol. 10, pp. 147-160.

9   **ISO** (1998) ISO 14041: Environmental management—life cycle assessment—goal and scope definition and inventory

10 **Herrmann, C., Betz, M., Kreissig, J., Itsubo, N.** (2000) Generic modules based LCA in electronics industry-a case study. In *Proceedings of the 2000 IEEE International Symposium on Electronics and the Environment,*pp. 124-129.

11 **Canadian National Office Of Pollution Prevention** (2000) Understanding the environmental aspects of electronic products: A life cycle assessment case study of a business telephone., pp. 3-4, 9-11, 40.

12 **Pollock, D., Coulon, R.** (1996) Life cycle assessment: of an inkjet print cartridge In *Proceedings of the 1996 IEEE International Symposium on Electronics and the Environment.* pp 154-158.

13 **Satake, K.** (1998) LCA for electronic products. A case study for CD-ROM drives. In *1998 IEEE International Symposium on Electronics and the Environment* pp 176 -179.

14 **Scheller, H., Hoffman, W.F.** (1999) Life cycle assessment of a telecommunication Product In *Proceedings First International Symposium on Environmentally Conscious Design and Inverse Manufacturing*, pp. 124-130.

15 **van der Wel, H., Reijnen, F.J.** (1997) Environmental evaluation of passive components with life cycle assessment In *Carts-Europe '97: 11th European Passive Components Symposium*, pp. 148.

16 **Ueno, T., Shiino, T., Onishi, H.** (1999) Evaluation of electronic components in life cycle assessment In *Journal of Cycles Waste Management*, Vol. 1, pp. 25-32.

17   **Taiariol, F., Fea, P., Papuzza, C., Casalino, R., Galbiati, E., Zappa, S.** (2001) Life cycle assessment of an integrated circuit product In *Proceedings of the 2001 IEEE International Symposium on Electronics and the Environment* pp.128 -133.

18   **Liu, J.** (2001) ACA bonding technology for low cost electronics packaging applications – current status and remaining challenges In *Soldering & Surface Mount Technology*, Vol. 13, pp. 39-47.

19   **Weidema, B.** (1998) Environmental Assessment of Products - A Textbook on LCA Helsinki, 3rd ed., pp. 14.

The recycling of telecommunication products – a case study in a Brazilian company

A C C MARQUES* and **R F CRUZ**
Department of Engineering, Furukawa Industrial S.A. Produtos Eletricos, Curitiba, Brazil
C CZIULIK
MSc Programme in Mechancial and Materials Engineering, Federal Centre for Technological Education, Curitiba, Brazil (*A C C Marques also of this establishment)

ABSTRACT

The environmental regulations are becoming stricter all over the world. Additionally, consumers are requiring that their products encompass environmentally friendly features and some governments are imposing them in order to trade goods. The Brazilian Parliament is debating a law project on environmental issues, which involves, amongst others, the "take back policy".

This paper presents a case study on a class of products used extensively in telecommunication field (optical and metallic cables). These products are manufactured by a leading company (Furukawa Industrial S.A.) in Brazil, which has recently been certified with the ISO 14001. The impacts of this kind of law on the products design and company's environmental strategy are discussed and based in this study, a preliminary solution for recycling these products is presented.

1   BRAZIL AND THE ENVIRONMENT

Brazil is a country located in South America, which area is 8.5 million km^2 and its population is about of 160 million inhabitants. Additionally, Brazil encompasses great part of the virgin forests in the world and 8% of the drinkable water available.

In the past few years, as everybody in the world, Brazilian people are becoming more concerned with environmental issues in the Earth. This awareness is mainly due to the access to information and because citizens are discovering their role in the society (i.e. as consumers, sales people, employees).

Furthermore, there are several companies in Brazil that are implementing leading actions concerning environmental issues. This occurs mainly because various organizations in the supply chain are being pressed from all sides to deliver environmentally compliant products. Thus, internal and global markets are signalling trends, imposing requirements and requesting products that cause less or no harm at all to the environment. More than a treat, environmental matters can provide a leading edge for those companies (and countries) that are capable of addressing them correctly.

1.1 Brazilian environmental legislation

From the approval, by the Brazilian Parliament, of the Law n° 9605, from February 12, 1998, (best known as Law for Ruling Crimes Against Environment), the awareness with environmental issues became an urgent topic to be considered by most of the Brazilian companies.

Additionally to this main law, there are several other regulations (approved by the federal, state and local regulatory bodies), that rule specifically the solid waste theme. As examples, there are:

- Ministerial Act n° 53, from March 1st, 1970, which regulates the solid waste disposal.
- State Law n° 12493, from January 22, 1999, which defines the criteria for the destination of solid waste, in Parana State.

1.2 Internal and external pressures

There are several kinds of pressure that are requesting from companies an efficient environmental management. A number of them are summarized in **Table 1**.

Table 1. Different kinds of topics that exert pressure, during the product development process, in Brazilian companies.

TOPIC	DESCRIPTION
Legislation	Constant reduction on the emission limits
Monetary Fines/ Costs	Vigorous checking by regulatory bodies and tendency on the rising of monetary fines
Personal Liability	Brazilian Law n ° 9605, from February 12, 1998
International Codes	Rio 92 – Agenda 21, Kyoto
Investors	Reports showing the eco-efficiency
Activists Organizations (NGO)	Increasing in the number of public protests
Customer Preferences	Changes in habits and preferences; eco-labels
Global Market	Adaptation to the International Codes

1.3 Brazilian national policy for solid waste

The Brazilian Parliament has been debating a Law Project concerning the Brazilian National Policy for Solid Waste. This Law establishes the "solidarity responsibility" for the management of solid waste. This means that everyone (*i.e.* consumers, companies, shops) is accountable for disposing the solid waste generated by the production and use of a specific item. However, in practice this Law places great amount of obligation on the Brazilian manufacturers and importers of machines, equipments, parts, components and packaging.

D002/018 © IMechE 2002

The great novelty of this Law is that the entrepreneurs become responsible for managing their products after the end of their useful life. Which means that companies will have to "take back" their products after they have been used by consumers.

In this way, when a optical cable or equipment assembled in a specific place, would have to be replaced, the company that have installed it, would be responsible to provide the adequate conditions to the client to remove, transport and make its final disposal.

It has already been foreseeing that this Law will have several modifications, mainly due to:

a) The costs allocated to the entrepreneurs;
b) Enormous logistic restrictions;
c) Difficulties in asking customers to return the products for the final disposal.

However, despite the alterations that will be implemented, this Law will have a great impact in the way companies are developing their products.

2 THE COMPANY

Furukawa is a pioneer and leader in several technological areas on a worldwide basis, being a member of a solid global group that encompasses, The Furukawa Electric Co., Ltd. (which has been widely operating in the power and communication markets since the early twenties), and Mitsui & Co., Ltd. (one of the major trading companies in the world).

Furukawa has been operating in Brazil since 1974, with excellent results both in the public and private markets. It is one of the first companies in its sector to implement an environmental certification. Because it stands on an industrial and technological forefront position it undertook the responsibility for the Environmental Management. As a large-size company, it has established as one of its priorities to supply its products with reduced pollution. Furukawa has adopted a policy in which its environmental goals meet the legal requirements, promote improvements and act in such a way that can prevent pollution. An Environmental Management System has been implemented with the purpose of continuously reducing the impact of the industrial activity on the environment described in the next section, by means of mainly improving the production processes.

Additionally, to protect the environment and contribute to the realization of a sustainable society, Furukawa is actively working to develop environment-friendly products and technologies. The aim is to develop a range of commercially viable "environment-harmonized" products at every stage, from materials selection, manufacture and use, to distribution and disposal (1).

3 CASE-STUDIES

The present case studies are limited to two of the main products (i.e. metallic and optical cables), designed and manufactured by Furukawa's Plant, located in Curitiba – Parana State. The focus of this examination is placed on illustrating several approaches that have been undertaken to address environmentally concerned issues in the following areas: design, manufacturing process and "take back". Additionally to that, it is presented an alternative approach that has been devised by another company (Resiways) that also aims at reducing the environmental impact with the remains of optical cables, nowadays disposed in a landfill.

3.1 Products description

The two types of cables analysed in this study are shown in **Figure 1**. The reasons behind this choice are that they involve the use of several different materials, mainly metals and plastics. Since, with certain care, almost any metal and much plastic can be recycled, an opportunity to implement a contribution in this field arises. However, plastic recycling faces one huge problem: plastic types must not be mixed for recycling and cannot be distinguished one type from another by sight or touch. It is well known that even a small amount of the wrong type of plastic can ruin the whole lot of melt. Thus, the main characteristics for both cables are given below:

3.1.1 Metallic cables

> **Uses**: telecommunications, data communication, broadband system (CATV).

> **Constructive characteristics**: The metallic cables, for telephone lines and electronic systems, are usually composed by a good conductor of electricity (in most cases, copper), insulated with a thermoplastic material. They are wounded in pairs (the most common case), triplets or four wires. Afterwards, they are joined together in a bundle, originating in this way, electronic cables (with 2 to 25 pairs), or telephonic cables (with 2 to 2400 pairs). Next, this bundle is protected with a plastic cover. However, in certain situations the core is shielded (using a metallic net, aluminium strip, copper strip, metallized polyester strip) against electromagnetic interference. Finally, the telephonic cables can be used, either: i/ externally: these form the net that connect cities, repetition stations and houses; ii/ internally: they are use inside a building and ensure the link between the stations and home devices. From the functional point of view and constructive characteristics both are very similar, although the ones to be used outside have a more robust design for protection against the elements, and those for internal use are built with non-flammable materials. **Figure 1** illustrates certain constructive characteristics of a metallic cable.

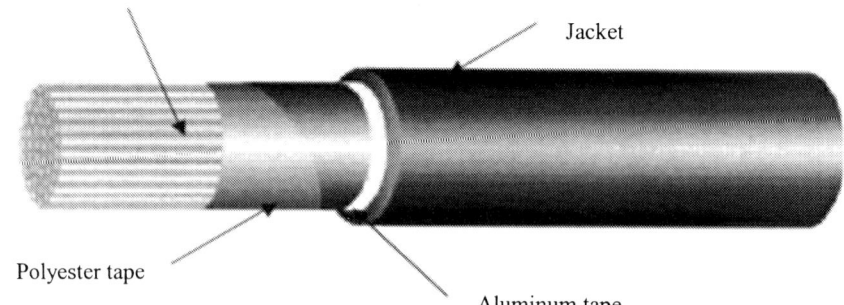

Copper with polyethylene insulation

Jacket

Polyester tape

Aluminum tape

Figure 1- Example of configuration in a metallic cable.

D002/018 © IMechE 2002

3.1.2 *Optical cables*

➤ **Uses**: telecommunications, data communication, broadband system (CATV).

➤ **Constructive characteristics**: in this case for loose tube cable, the main elements are polyethylene jacket, aramide, water-blocking tapes, Polybutylene Terephtalate – PBT tubes, Fibber Reinforced Pultruded - FRP, jelly and optical fibber (glass and resins). The same optical fibber can be use in several different environments, which does not apply to the whole cable. The cables are designed to withstand particular conditions, enabling an adequate and proper environment for the optical fibbers inside them. Therefore, the design of a cable will have to consider the environmental conditions where they are going to be installed. The main environments for optical cables can be classified as follows: i/ internal devices (e.g. inside a telephonic switching system or a computer); ii/ inside buildings (e.g. between walls or above ceilings, typically between distribution nodes in a building); iii/ aerial cables (e.g. hanging on external posts. In this case, they can be supported by specific metallic cables); iv/ self supported cables, completely dielectric; v/ optical cables installed inside a buried plastic duct; vi/ buried optical cables (e.g. cables placed directly inside a trench); vii/ submarine optical cables (e.g. submerged in salted or sweet water); viii/ hybrid cables for energy and optical signals. **Figure 2** shows a typical configuration of an optical cable.

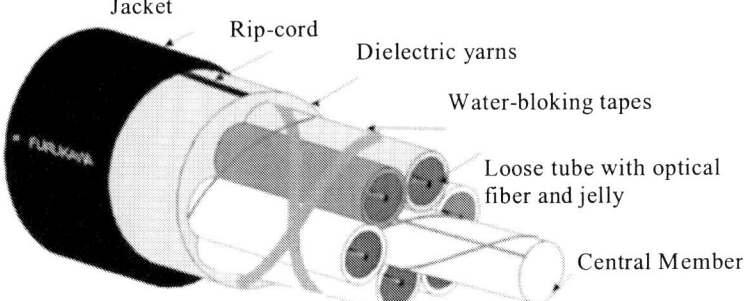

Figure 2- Example of configuration in an optical cable

Figure 3, allows to contrast the constructive characteristics of both cable types (metallic and optical).

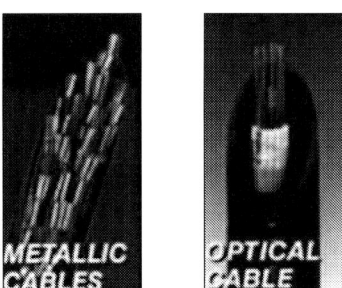

Figure 3. Examples of metallic and optical cables.

3.2 Implemented actions at Furukawa

This section contains a summary of actions that have been already been taken into account, aiming at experimenting some strategies

3.2.1 Metallic cables

Table 2 presents specific implementations aiming at reducing the impact on the environment, concerning at metallic cables.

Table 2. Implementations – metallic cables

AREA	ACTION	MAIN BENEFITS
Design	✓ *Employment of:* →*Materials without heavy metals;* ✓ → *Materials LSZH (Low Smoke Zero Halogen).*	*This environment-harmonized materials can minimize the environmental effects at landfill and incineration. Moreover, since the material does not contains toxic heavy metals without mentioning halogens, it can cope with recycling processes, thus making it very useful for using in cables (2).*
Manufacturing Process	✓ *Re-use of jacket materials;* ✓ *Recycling of cables.*	*- Reduction of the consumption of raw materials;* *- Financial benefit from selling those materials (Cooper and polymers) separately.*
Take Back	✓ *Analysis focusing the recycling of cables.*	*- Financial benefit from selling those materials (Cooper and polymers) separately.*

Analysis:

a) The products with reduced environmental impact do not emit toxic by-products when they are eventually disposed of by incineration or in landfills, thus reducing environmental impact. To develop products that reuse waste materials, products that feature unification of materials to facilitate recycling and thus do not leave residual waste products *(1)*.

b) The re-use of jacket materials has been previously very carefully studied, in order to maintain all needed properties for a metallic cable (e.g. resistance to UV rays). Thus, the jacket materials can be re-used (in an adequate proportion with virgin materials), in such a way that the final product yields similar properties to those manufactured with raw materials.

c) For recycling the metallic cables the following procedure has been devised to separate the different types of material: the first step is to remove manually the cable's cover (polyethylene), using the existing ripcord. Next, again through a manual process, the aluminium strip is separated. Finally, the individual wires that remain are placed is a purpose-built equipment that crushes them. Using an intense airflow and specific properties of each material, the cooper is isolated from the insulation (i.e.

D002/018 © IMechE 2002

polyethylene, polyvinyl chloride and nylon). Then the copper is sold to be recouped and the insulation is sold to be transformed at another products like hoses.

d) The analysis focusing recycling of cables has been conducted with twelve more representative metallic cables manufactured by Furukawa. In this study, were considered the following aspects:

- The materials of the components that integrate a cable, which have a commercial value as a scrap material are: cooper, aluminium and polyethylene.

- During the separation process, the accounted costs were: the machine costs and work force costs involved in the removal the plastic, aluminium strip; the costs of separation process of the cooper and plastic cover; the costs with electric energy. The main incomes were derived from the selling of individual materials (cooper, aluminium and polyethylene).

For the sake of illustration, there are given some figures for a metallic cable with 900 pairs of cooper conductors. For this cable, the financial benefit (incomes - costs) represented about 35% of the cable total cost. Therefore, according to this preliminary analysis and from the financial point of view, the "take back" approach for metallic cables would not represent a major problem for Furukawa, in case of the approval by the Brazilian Parliament of the Brazilian National Policy for Solid Waste. It is important to emphasize that it is being already analysed a request from a major telephone company which is asking to exchange certain quantity of metallic cable scrap (in various lengths) for specific cables with determined lengths. In this analysis the commercial value of the cable scrap is exchanged by new cables. In this way, the take back approach would have its economic value based on the commercial significance of the scrap parts.

3.2.2 Optical cables

Table 3 presents specific implementations aiming at reducing the impact on the environment, concerning at optical cables.

Table 3. Implementations – optical cables

AREA	ACTION	MAIN BENEFITS
Design	✓ Materials without heavy metals; ✓ Materials LSZH (Low Smoke Zero Halogen).	- The environment-harmonized materials can minimize the environmental effects at landfill and incineration. Moreover, since the material does not contains toxic heavy metals without mentioning halogens, it can cope with recycling processes, thus making it very useful for using in cables (2).
Manufacturing Process	✓ Re-use of jacket materials; ✓ Disposing the materials in the landfill.	- Reduction of the consumption of raw materials.
Take Back	✓ Not yet implemented.	-

Analysis:

a) The considerations regarding the use of environment-harmonized materials and the re-use of jacket parts that come from optical cables are identical to those described in the section (when discussing metallic cables).

b) Differently form metallic cables, optical cables are nowadays disposed in a landfill. Thus, there is a financial cost to implement this disposal, since Furukawa has to pay for the service of disposing optical cables in a specific place.

c) Regarding the take back approach, two different options will be examined section 3.3.

3.2.3 General comments on Furukawa's approach

Despite not having being used explicitly, it is possible to realize that the measures implemented at Furukawa are complying with the Design for Environment Guidelines, proposed by Billatos and Nevrekar *(3)*. Those guidelines more closely matched are describe in **Table 4**.

Table 4. DFE Guidelines according to (1).

GUIDELINE Nº	DESCRIPTION	MAPPING TO FURUKAWA
2	→ Find multiple or secondary uses for a product. Disposal will be less of a problem if a product has more "intrinsic value".	→ Re-use of jacket materials, at the manufacturing stage of metallic and optical cables.
3	→ To ensure easier recycling, use materials that match each other closely or the same material. Look for ways to use recycled materials as starting compounds for a product.	→ Recycling of copper and polymers (metallic cables), at the manufacturing stage at Furukawa's plant. → Initiative developed by Resiways in order to recycle optical cables.
8	→ Establish a network of producers and suppliers to form the beginnings of the "industrial ecosystem" and facilitate DFE efforts.	→ Contacts established with suppliers of raw materials.
9	→ Look to reduce waste by-products streams in manufacturing processes. Seek out non-hazardous solvents and cleaning materials. Reduce energy consumption by eliminating unnecessary manufacturing steps.	→ Specification of non-hazardous materials at the design stage of metallic and optical cables.

This shows that adopting certain measures, companies can comply with recommended guidelines, which can help in contributing to match International Standards.

3.3 Alternative approaches for optical cables

The costs associated with separation and purification of glass and plastic components in a fibber cable far exceeds the manufacturing/processing costs of the virgin materials. Unlike

D002/018 © IMechE 2002

copper cables, where the price of copper helps drive the recycling business, there is no high-value material easily reclaimed from the fibber-cable *(4)*.

Thus, this section presents two alternative approaches towards recycling optical cables. The first one has been undertaken by an entrepreneur that has envisaged the possibility of using optical cables material as one of the components for manufacturing moulded products. The next alternative, it is an option described in the literature that the authors are suggesting to be studied with more detail by the Brazilian manufacturers in order to give a final destination to optical cables remains.

3.3.1 A novel material

This process has been developed and patented by a private company, called Resiways *(5)*. The development of this new recycling technology, it is a novel and important approach to control the pollution generated by plastic and wood waste, which until some years ago were difficult to recycle. The whole process consists of chopping the optical cables and mixing them with wood chips. The next step, involves the fusion of plastic parts with wood, producing an amorphous mass. This new material can be used, via moulding techniques, for manufacturing several products with applications in different market segments. Examples of new products that can be obtained with this recycled material are shown in **Table 5**. Furthermore, **Figure 4** contains an example of a product that can be obtained with this new material.

Table 5. Different uses for a recycled material obtained from optical cables scrap.

FIELD OF APPLICATION	DESCRIPTION
Furniture	*Backs and seats for chairs; school seats; furniture parts and reinforcements.*
Transport and Packing	*Pallets; reels for wires and cables, boxes.*
Automotive industry	*Sides; roofs; boot cover; sun break devices; internal floor for buses and trucks.*
Civil Construction	*Roofs; floors; structural components; protective devices.*

Figure 4. Example of product (reel) obtained with recycled optical cables.

It has to be said that the reel, after the end of its useful life, can also be reprocessed and employed as a compound material for moulding other products.

3.3.2 Cement production

This approach has been found in the literature (6). The authors envisaged this alternative because an optical fibber-cable is composed of plastics and other materials such as Fe, Al and SiO_2. Since the main components of cement are CaO, SiO_2, Al_2O_3 and Fe_2O_3, an opportunity to incorporate Fe, Al and SiO_2 as cement components arises. Moreover, those plastics parts can be used as heat source for cement calcinations.

The poisonous gases when burning optical cables, decompose during the cement production process because of the high calcinations temperatures (around 2000 $^{\circ}$.C). Therefore, recycling in the cement production process does not generate more waste or poisonous gases.

Until now, there is no information about any company implementing such approach in Brazil. However, since Brazil occupies an important position in the world market cement production it is possible to anticipate that this alternative can become very popular in the near future.

4   FINAL REMARKS

This study has shown that companies have to become more aware of the environmental issues due to legislation and social responsibilities.

Additionally, it has been perceived that ISO 14000 just supplies the guidelines to implement a more environmentally oriented approach to develop products and business. In this paper, it has been presented an internal initiative from Furukawa that is looking ahead in its way of designing, manufacturing and, if it is the case, recycling the products that exist in its portfolio. Here, it has been examined the approaches that have been undertaken towards metallic and optical cables. However, it is expected that the whole company can benefit from this preliminary study. Thus, for recycling metallic cable scrap it is important to emphasize that companies, which can recycle this products, can also obtain a competitive advantage by exchanging these scraps for new cables.

The alternative approaches (from Resiways and cement production) have portrayed feasible options that have already been translated into a product with reduced environmental impact. However, it is very likely that other methods for recycling metallic and optical cables are going to be devised. To define a preferable strategy for end-of-life treatment, the literature available (7) suggests that the following issues should be considered:

a) Does implementing the take-back legislation affect the choice of strategy?

b) Do customers have preferences regarding the issue?

c) What is the most beneficial aspect from an economic perspective?

d) What option is preferable from an environmental standpoint?

Finally, it is important to emphasize that companies have to be more pro-actives, not only responding to market pressures, but also anticipating demands and effectively illustrating how they are collaborating towards preserving the environment.

5   CONCLUSION

The solid waste policy, as soon as, is approved by the Brazilian Parliament will have a great impact on the product development strategies.

The product development process will have to address these issues, and produce innovative solutions, which are cost effective and environment-friendly.

On the other hand, companies that introduce this approach, as soon as possible, can obtain a competitive advantage, showing its concerns and results to the customers and competitors.

This work has presented effective measures that have been implemented at design and manufacturing stages by Furukawa, a company that is already very much concerned with environmental issues and its consequences. Additionally, the initiative proposed by Resiways has proved to be a valid contribution towards reducing the environmental impact, when disposing optical cables.

6   ACKNOWLEDGMENTS

Mr. Jamil Azrak (Resiways) is thanked for supplying data for this paper.

REFERENCES

1   **Furukawa Electric.** Environment Report 2001.

2   **Yamada, H. and Nishiguchi, M.** (2000). Development of Environment-Harmonized Electric Wires. Furukawa Review, n°19, pp. 81-84.

3   **Billatos, S. B. and Nevrekar, V. V.** (1994). Challenges and Practical Solutions to Designing for the Environment. Design for Manufacturability – 1994. National Design Engineering Conference, Chicago, IL, USA, Mar. 19-23. Vol. 67. New York: ASME. pp. 49-76.

4   **Bowner, T.** (1997). Environmentally Friendly Strategies for Telecommunications Providers. Proceedings of 46th International Wire & Cable Symposium, pp. 426-435.

5   **Resiways.** Um Inédito Processo de Reciclagem. São Paulo, 2001. <http://sites.uol.com.br/resiways> (accessed in April 14th, 2002).

6   **Konaka, T., Shima, S., Hara, T., Nishi, S.** and **Imoto, I.** (1998). Optical Cable Recycling. Proceedings of 47th International Wire & Cable Symposium, pp. 740-744.

7   **Furuhjelm, J., Yasuda, Y., Trankell, R.** Recycling of Telecommunication Products in Europe, Japan and USA.

Telephone re-manufacture – sustainable development in action

K SNOWDON
Nortel Networks plc, UK
C BACKHOUSE and **A CLEGG**
Wolfson School of Mechanical and Manufacturing Engineering, Loughborough University, UK

Abstract

Commercial and legislative pressures are forcing Electronics Company's worldwide to adopt a total life cycle approach to the design and manufacture of their products. Key to sustainable development in this sector, is the ability to re-manufacture, upgrade and repair old products. Current processes and design pointers in the sustainable manufacture of telephone handsets, key switch units and fixed wireless access products will be described. The impact of the lessons learned in the teaching of sustainability to new undergraduate designers and manufacturing engineers will be discussed.

1. INTRODUCTION.

During the past 40 years advances in electronics have revolutionised the way in which we live and do business. Sophisticated hardware and software permits voice communication virtually anywhere in the world and brings video-on-demand to the home via the Internet. The electronic content of cars now reaches in excess of 25% of their costs and CAT (Computer Aided Tomography) scanners, heart pacemakers and medical diagnostics improve the health of people everywhere.

In 1998 (1) the ICT (Information and Communications Technology) Industry in Western Europe had a turnover in excess of 392 billion euros with telecommunications services and equipment accounting for 51% of this expenditure. However, every technological advance brings with it new demands on the Earth's resources. The ICT industry worldwide must increasingly make its products and services "sustainable" to ensure that future generations will enjoy the same quality of life as today's citizens.

The concept of sustainability or eco-efficiency, was born at the 1992 "Earth Summit". Since 1992 the World Business Council for Sustainable Development (WBCSD) has done much to develop the concept into a meaningful principle which business can readily adopt.

WBCSD has identified seven steps towards eco-efficiency and hence sustainable development;

1. Reduce the material intensity of goods and services.
2. Reduce the energy intensity of goods and services.
3. Reduce toxic dispersion.

4. Maximise sustainable use of renewable resources.
5. Enhance material recyclability.
6. Extend product durability.
7. Increase the service intensity of goods and services.

Life Cycle Assessment (2) of telecommunications equipment has shown that the manufacture and use life cycle stages often have the biggest impacts on the Environment. This paper will address design aspects in telecommunications equipment re- manufacturing concerning five of the above topics viz:

Reduce the material intensity of goods and services.
Reduce toxic dispersion.
Maximise sustainable use of renewable resources.
Enhance material recyclability.
Extend product durability.

2. REDUCED MATERIALS INTENSITY.

 The environmental impact of a product can often be reduced by the process of dematerialisation, which seeks to reduce the volume and weight of materials used in the manufacture of the product. The drive to do this already exists as a way of reducing costs. One example of this, is the reduction in the size of PCBs (Printed Circuit Boards) and individual components. Boards have reduced in size by some 75% over the last 10 years whilst at the same time the functions delivered by the board have increased dramatically. Even if it is not possible to recover and reuse the board or its component parts, the amount of material used in their production has decreased and so has the amount of material consigned to waste at end of life. See figure1 below showing telephone PCBs 10, 7, 4 and 1 year old respectively from left to right.

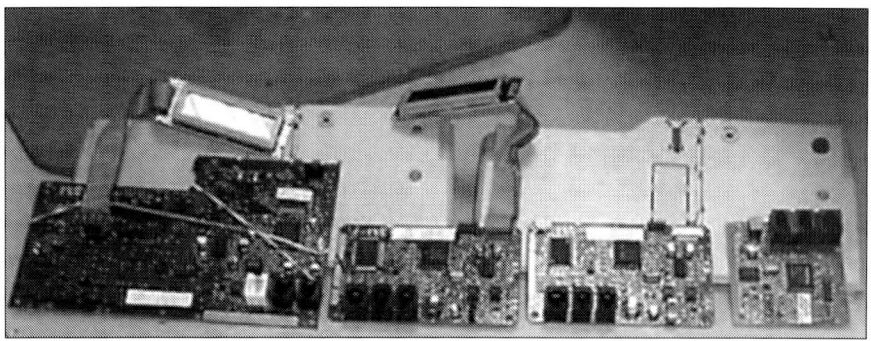

Figure 1. PCBs from telephones 10,7, 4 and 1 years old.

3. REDUCED TOXIC DISPERSION.

The telecommunications industry deploys many complex materials and processes in the manufacture, servicing and use of its communications equipment. Recent legislation such as the draft European RoHS (Restriction of certain Hazardous Substances) Directive has stimulated a lot of research into hazardous materials replacement for lead in electronic interconnect . Lead containing interconnect is likely to be banned in electronic equipment for sale in Europe after 1/1/2006.

Worries of lead leaching out of old electronic equipment (3) into water supplies and resultant poisoning of humans and the eco-system are the major reason for a ban on lead materials in the RoHS draft directive. Lead based materials are used in the electronics industry in TV cathode ray tubes, interconnect on Printed Circuit Boards (PCBs) and in certain components such as ceramic capacitors, piezoelectric devices and ferroelectric displays.

The development of lead-free interconnect requires the replacement of lead (4-7) in three key areas of a solder joint- the solder, the finish on the PCB and the finish on the components legs. See figure 2 below.

Figure 2, A lead-free solder joint

As can be seen in table,1 there are many candidates for Pb- free solder and Pb- free PCB finishes. However, there is no single drop in replacement for conventional eutectic tin-lead solder.

Table 1: Lead -free interconnect selections

Solder Alloy	PCB Finishes	Component Finishes
99.3Sn,0.7Cu	Organic	Sn
97.5Sn,2.5Ag	Organo-silver	Pd/Ni
SnCuAg	Immersion Au/Ni	Au
SnBi	HASL Sn/Cu	Ag
SnAgBi		Sn/Bi
SnSb		Ni

In 1996, Nortel Networks manufactured the world's first virtually lead-free telephone (8) using a lead-free, tin-copper solder, an organic PCB finish and as many lead free components as could procured be at the time. The resulting telephone PCB showed a x11 reduction in the amount of lead leachate after simulated end-of- life testing.

4. MAXIMISE USE OF SUSTAINABLE RESOURCES.

Telephone units may be returned to Nortel in individual cardboard packages. Cardboard packaging is made from a renewable resource-wood. If these are simply discarded they represent a waste of the materials and money. Nortel uses these boxes to return the refurbished units to the customer. The company places a recycling sticker on the box , which can be ticked to show the number of times the box is transferred between the company and the customer. The current record for such return journeys is 7. See figure 3 overleaf showing several cardboard boxes ready for re-use.

Individually boxed sets are packed into larger transportation boxes. Again, the boxes that come from customers are recycled and used to return the refurbished product back to the customer. Such boxes may also see several repeat journeys.

Packaging has several impacts on the environment and is the subject of European Union legislation regarding its use and disposal. The redesign of standard packaging can have environmental advantages on several fronts. For example, a standard Nortel transportation carton use to hold 20 boxed sets. By re-designing the telephone set box to occupy a smaller volume the number of boxed sets per carton can be increased to 30. Apart from the significant reduction of cardboard in the set box there is an environmental advantage in the reduction of transport journeys necessary because more sets are transported per lorry load

D002/013 © IMechE 2002

Figure 3. Cardboard boxes labelled up for re-use for packaging re-furbished telephones.

5. ENHANCED MATERIAL RECYCLABILITY-PLASTICS RECYCLING.

The materials composition by weight of a typical Nortel telephone, packaged ready for delivery is shown below in table 2.

Table 2. The composition by weight, of a typical Nortel telephone

Material		Weight in Grammes	Percentage
ABS	Plastic	769	52.13%
Acrylic (Lens)		17	1.15%
Acetal (Pivots)		4.5	0.31%
Polyester (Flexi circuit)		2.5	0.17%

Polycarbonate (Flexible windows)	2	0.14%
Synthetic Rubber	1	0.07%
Silicone Rubber	27	1.83%
Foam	0.25	0.02%
Paper	68	4.61%
Polythene	7.5	0.51%
LCD.	26	1.76%
Acoustic devices (Inc. steel, magnets, paper etc)	159	10.78%
PCB	51	3.46%
PCB Components	68	4.61%
Steel (Zinc plated)	21	1.42%
Cordage & wiring	97.5	6.61%
Carton	154	10.44%
(Packaging		

By inspection of table 2 above, it can be seen that plastics account for > 54% by weight of the telephone and therefore should be attractive from a materials recycling standpoint. However, it should also be noted that there are at least six different plastics present and separation of these per material type is needed for efficient recycling. Contamination of one plastic by another, especially PVC and/or acetal- by only a small amount, can lead to the contaminated material being worthless.

Nortel has therefore replaced acetal plastic key caps with the much more compatible PBT (Poybutylene Terephthalate) in its more recent models.

Another factor, which aids plastics recycling, is that of materials identification. Permanent marking of a moulding by laser or in-mould engraving to international standards such as ISO 11469 greatly enhances the possibility of recycling.

Finding uses for recycled, engineering plastics is however, much more difficult. Concerns remain about quality, colour consistency and stability, availability in bulk multi-tonne lots, continuity of supply and single source supply agreements.

Good eco-efficient design for plastics recycling includes;

- Use the smallest number of different types of plastics possible
- Permanently identify large plastic parts > 50 gm in weight to ISO 11469 standard
- Use as few different colours as possible
- Develop closed loop recycling between plastics supplier, manufacturer and user.

6 EXTENSION OF PRODUCT DURABILITY.

Nortel Networks has developed a product hierarchy in terms of economic and eco-efficiency as durability options. These are especially applicable to the Key Switch Units; see figure 5 overleaf.

D002/013 © IMechE 2002

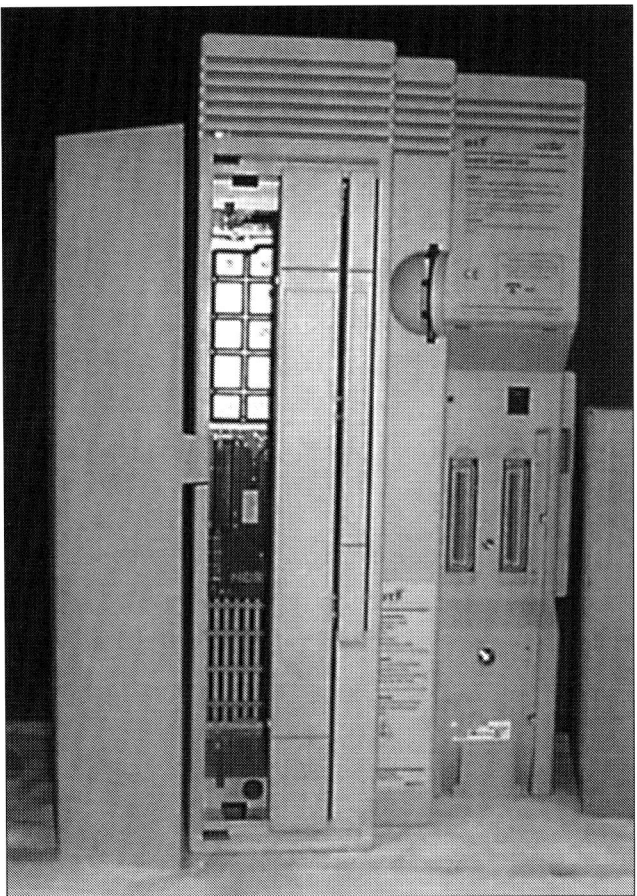

Figure 4. Key switch units used to connect internal phones with outside lines.

1) Software upgrades to existing equipment. This can save the production of new hardware, which can often weigh in excess of 30 kilograms per unit.
2) Hardware upgrades to existing equipment. e.g. swapping PCBs with more advanced features. Again re-use of >90% by weight, of old hardware can be realised.
3) Repair of consumable components e.g. integrated circuits, buttons, switches, cables etc.
4) Cleaning and/or painting of the ABS plastic consoles.

7. COMMUNICATING THE LESSONS OF SUSTAINABLE DESIGN.

As part of the Royal Academy of Engineering Sustainable development initiative, one of the authors became a visiting professor to the Mechanical and Manufacturing Engineering department of the Wolfson School at Loughborough University in 1999. The purpose of this appointment was to transfer industrial experience to the new generation of designers and manufacturing engineers. The following section describes the procedure and experiences from this technology transfer.

All first year students at Loughborough University participate in the Integrating Studies programme that enables students to explore manufacturing processes and materials requirements. The programme encompasses several activities that are particularly suited to the inclusion of environmental issues. Each student is required to produce a technical report on a subject related to manufacturing. The topic for the report is decided following discussion between the student and his/her personal tutor. Each tutor is circulated with a list of environmental topics from which the student can make a selection. However, environmental topics are not mandatory.

Students also work on 'dismantle and discover' group projects in which the principal objective is to encourage students to make the links between their academic study of materials and manufacturing processes by studying real products. To introduce an environmental context, the students were encouraged to consider the ease with which products could be dismantled and reusable or recyclable elements could be extracted. The final element of the Integrating Studies module is the Case Study. Invited Guests invited from industry present a case study, in this instance telephone remanufacture at Nortel Networks.

The time available to present the case study was 2 hours, the class size was 86 and the activity was presented in a tiered lecture theatre. The activity was not formally assessed. The objectives of the case study were:

To identify the environmental impacts of a manufacturing company.
To identify the actions of governments to encourage companies to reduce such impacts.
To identify Nortel's response with emphasis on the Waste Electronic and Electrical Equipment Directive and telephone remanufacture.

The case study required that the students be actively involved. Following a brief introduction, the students were invited to work in small groups to identify, in general terms, the environmental impacts that a typical manufacturing company might have. The students' views were elicited and used to produce a composite list. This was followed by a similar activity that sought their views on the role of governments in promoting industrial environmental responsibility. To follow this, the Industrial Professor presented an overview of Nortel Networks in the context of the impacts of the telecommunications and electronic industries on the environment.

The central activity of the case study was the practical disassembly that required the students to dismantle telephones and consoles so that they could comment on the design and manufacturing aspects of the remanufacture of such equipment. The students were asked to report back on their findings. To complete the session the Industrial Professor outlined the experiences of Nortel Networks/Solectron at their Cwmcarn remanufacturing site. A typical workbench from the factory is shown below in figure 5.

Figure 5. A work bench from the Cwmcarn factory.

The students were asked to provide feedback on the activity. This was specific to the case study because the feedback assesses the module in respect of the range of activities it encompasses. The students were asked to score from 1 to 5, the four aspects shown overleaf in table 3.

Table 3: Case study feedback response

Aspect	Score
New knowledge gained	3.28
Intellectual challenge	2.78
Interest value	3.76
Enjoyment value	3.98

The feedback was generally positive and the students enjoyed the activity (9). The lower scores for new knowledge gained and intellectual challenge suggest that the lecturers underestimated their audience. The case study material is being incorporated in a web site for future groups to access. This will reduce the need for the formal presentation of factual information within the case study activity.

8. CONCLUSIONS.

Design changes in materials selection, product miniaturization, toxic materials reduction, re-use of renewable transport packaging and software and hardware upgrades have enabled equipment re-manufacture to have a sustainable future in telecommunications equipment production..

The lessons learnt from this experience are now being incorporated into teaching material for the next generation of undergraduate designers and manufacturing engineers.

ACKNOWLEDGEMENTS.

The authors would like to express their thanks to the Royal Academy of Engineering for their financial support and encouragement in the production of this paper in particular Dr David Foxley, Manager, Engineering Design Education and the Directors of Nortel Networks and Solectron UK for the supply of refurbished equipment for study.

REFERENCES.

(1) European Information Technology Observatory (EITO) 1999.
(2) S.G.Anders et al, Journal of Electronics Manufacturing, Vol 10, No3 (2000), p147
(3)II.K.Charles Jr and N. Sinnadurai ,"Microelectronics; Rising to the Environmental Challenge" IMAPS 12th European Conference, Harrogate UK, June 1999, p 78.

(4) C. Melton "Wettability of lead-free solders" Journal of Materials, July 1993, p33

(5)J H Vincent and G Humphries " Lead-free solders", GEC Journal of Research, Vol 11, No2, 1994.

(6)"Matsushita Develop a Lead-free Solder",MCC Global Technology Monitor, March/April 1996 p56

(7)B. Craig "Lead-Free solders, an up-date",Electronic Production, July 1994, p21.

(8)K G Snowdon, C G Tanner et al "Lead-free soldering-Applications of Lead Free Solders in the Telecommunications Industry" Future Circuits International June 1998, p 21.

(9)Snowdon, K.G., Clegg, A.J. and Backhouse, C.J., Educating engineers in engineering design for sustainable development. *Proc. Electronics Goes Green 2000+.* Berlin, Germany, 841-844 (2000).

Educational Issues

A case study approach to the teaching of design for sustainability – the Royal Academy of Engineering scheme for visiting professors

P NORMAN, S HAILE, and **J HULSE**
Department of Chemical and Process Engineering, University of Newcastle-upon-Tyne, UK

Abstract

The Royal Academy of Engineering supports Visiting Professors in Engineering Design for Sustainable Development in fifteen universities. This paper describes the experiences gained over the first three years of the scheme at Newcastle. A number of case studies have been completed and introduced into the curriculum in several Departments at both undergraduate and Masters levels. The case studies cover: *product design, social aspects, decision making, systems specification* and *sustainable transport.* The scheme has been extended for a further two years to allow concentration on dissemination, a major objective being to make the case study material available in a form suitable for Internet based distance learning.

1 INTRODUCTION

The Royal Academy of Engineering received a grant-in-aid from the government's Office of Science and Technology for an initial 3-year period (1998-2001) to establish a scheme for the placement of Visiting Professors (VPs) in Engineering Design for Sustainable Development in selected institutions of Higher Education. This provided funding of £20k per year for three years to support each appointment. Over the period, appointments have been made at the rate of five per year awarded on the basis of proposals received from universities invited to bid each year. As part of the selection process, the Academy identified a number of themes with a strong emphasis on their multidisciplinary character (e.g. built environment, materials and manufacturing, etc.) across the whole field of engineering. The bid from Newcastle was co-ordinated by the Faculty of Engineering's Engineering Design Centre (EDC) in conjunction with the Department of Chemical and Process Engineering.

Because engineering decisions can have a major impact on society on all time scales, it is imperative to enable tomorrow's engineers by providing them with the tools and methodologies needed to address the problems of sustainable development. The aim of the Academy's scheme is to raise the awareness of engineers of the core issues of sustainable development and the achievement of a sustainable society, thereby enabling them to take a holistic view of society's needs(1). One approach to the realisation of this aim is to make available good quality teaching material which enhances both the understanding and the practice of Sustainable Development. The Academy's view was that this could best be achieved by means of leading industrial practitioners working in unison with experienced teachers. The new scheme therefore needed to be based on a very close and resource intensive partnership between industry and academia.

2 OBJECTIVES OF THE ROYAL ACADEMY OF ENGINEERING VISITING PROFESSORS SCHEME

A selection of objectives published by the Academy is summarised below:

1. To generate a knowledgebase of case studies of decisions on a diverse range of developments with an engineering dimension, and definable links with Sustainable Development indicators
2. To identify elements of good practice in engineering design
3. To disseminate case study material into the university system
4. To assist with the incorporation of general principles into the engineering curriculum.

Some of the ways this was to be achieved include:

1. Securing the appointment of engineers with experience of how decisions are made as VPs
2. Achieving a balanced distribution of resource across all fields of engineering
3. Facilitating communication between and co-ordination of the work of the VPs during their tenure
4. Engaging other bodies in enabling universities to adapt the curriculum to take advantage of the inputs from the scheme
5. Establishing and maintaining liaison with Academies and organisations in other countries with similar interests.

To this we would add the requirement to engage fully with the academic champions in each university and encourage them to disseminate the outcomes to their colleagues and the wider academic community.

3 PROGRESS OF THE SCHEME AT NEWCASTLE

The scheme started at Newcastle in November 1998 with the appointment of Jeff Hulse, a senior manager with BASF, as the Visiting Professor. It has resulted in the production of a number of case studies largely, but not exclusively, with a process engineering flavour. The Academy has now extended the resourcing of the Newcastle programme for a further two years with the aim of concentrating on dissemination, although new case studies are still being actively developed. The Academy's original brief has been interpreted in a variety of ways in the participating institutions and each has had considerable freedom in the way the resources provided have been used. At Newcastle, the original aims were stated as:

1. Familiarisation with existing University teaching programmes involving environment and sustainability.
2. Reviews of current practice in design for sustainable development in the made-to-order industries, especially process engineering.
3. Discussions with Faculty academic staff to facilitate the production of recommendations for enhanced teaching of sustainable development.
4. Assembly and development of case study material.
5. Development of material for dissemination to a wider audience.
6. On going programme of academic evaluation of case study material together with exploration and development of delivery mechanisms.

D002/009 © IMechE 2002

This was refined at the start of the programme into four areas of activity:
1. Networking with providers of teaching of sustainability, and with industrial practitioners
2. Benchmarking
3. Assembly of resources
4. Development of teaching materials

The programme is managed by two academics from the Department of Chemical and Process Engineering – Peter Norman and Sue Haile.

3.1 Networking activities

A major aim in appointing VPs was to capitalise on the wealth of experience in sustainability that is accumulating both in industry and academia. This was to be achieved by engaging respected, high profile industrial practitioners and bringing them into contact with the academic environment. During the three years of his appointment Jeff Hulse has visited a number of sites and had discussions with individuals on sustainability issues. His involvement with the Chemical Industries Association as an advisor, and his work with the Teesside clean air programme has provided invaluable input to the academics. This networking activity is regularly reported back to the programme team. As well as industrial contacts, Jeff has met with fellow Academy VPs, as well as academics at other institutions. A small sub-group of VPs and academic staff from Chemical Engineering Departments at Aston, Oxford, Surrey and Newcastle have met regularly to discuss progress and to exchange ideas. Jeff has also been involved with the working party set up by the Institution of Chemical Engineers to look at the teaching of sustainability.

The programme team has had some success in crossing disciplinary boundaries. The Department's MSc in Clean Technology brings together engineers and scientists from a range of disciplines in addition to Chemical Engineering and the case studies that have been developed have exploited this diversity very successfully. Notably, there has been interest from colleagues in both Mechanical and Civil Engineering in a case study which is aimed at sustainability issues relating to the selection of motor vehicle systems. There is an automotive MEng at Newcastle and some of the students from that course attend a Sustainable Engineering module which makes use of the motor vehicle case study. Students from other degree courses such as Marine Technology, Materials Engineering and Manufacturing have also had the opportunity to attend case study workshops, some of these being Erasmus students from other EU countries.

While many academic colleagues recognise the relevance of sustainability to their subject areas and are willing to include it in their modules, a major issue is the squeeze on an already tight curriculum. However, the very nature of sustainability lends itself to demonstration in a wide variety of course topics. Consequently, smaller, simpler case studies are being worked up that can be used as examples of sustainability in action in specific topic areas such as refrigeration, sludge disposal, heat utilisation and material separation processes. However, one subject area that is amenable to direct exposition of sustainability principles is Design. In Chemical Engineering at Newcastle there are specific modules on Design in each Stage of both the BEng and MEng courses and these provide excellent opportunities for case study based teaching of design for sustainability.

3.2 Benchmarking

The general knowledge of sustainability issues is being monitored in each of the three years of the programme. This activity has involved first and final year Chemical Engineering undergraduates as well as Clean Technology MSc students. This allows the team to benchmark progress in raising awareness in students of sustainability issues. The data is presented as a bar chart showing the relative response regarding knowledge about sustainability. The responses are grouped into the same categories as those used in the 'Toyne II' Report(2)

GI = Global issues
KSC = Key sustainability concepts
LPC = Legislation, policy and control
T&T = Tools and techniques
R = Responsibility
K = Key skills

It is interesting to compare the knowledge claimed by first year undergraduates and that claimed by Clean Technology MSc students. The latter mainly comprise graduate chemical engineers, environmental scientists, chemists and biologists. While the former clearly have little idea across the whole range of issues, the MSc students demonstrate significant improvement on general issues and the tools and techniques of sustainable development, presumably as a result of the instruction they have received. However, some important topics were identified as being unfamiliar at the MSc level e.g. Product Stewardship, and Social issues such as social reporting. These topics were integrated into the necessary underpinning lectures given prior to the case study exercises. The key conclusions showed the need to introduce more social aspects e.g. Risk communication and Ethical issues.

D002/009 © IMechE 2002

1st Year Chemical Engineers 1999

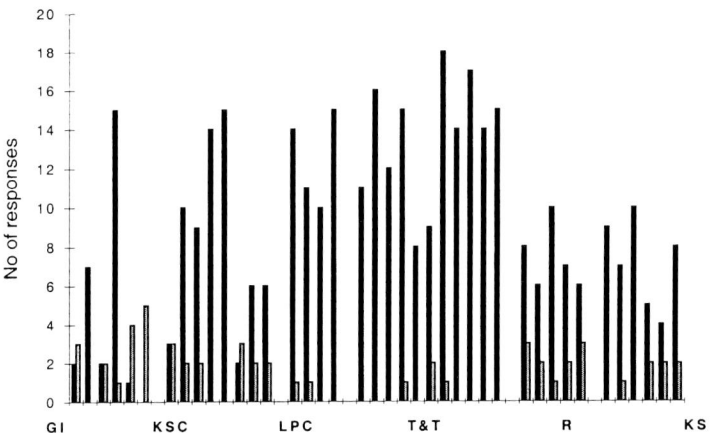

MSc Clean Technology end 1999

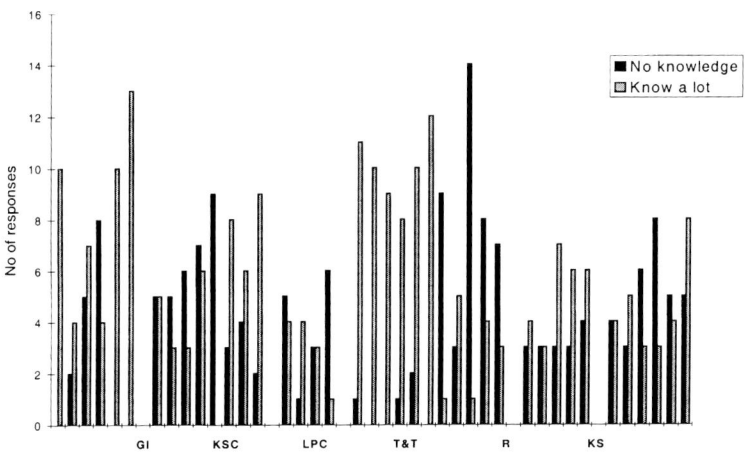

3.3 Resources

A collection of reference material is being assembled (Books, reports, articles, papers, etc.) which are available as a resource for both students and staff. In addition, an intended deliverable at the end of the programme will be a *Resource Pack*, containing sources of information related to sustainable development.

3.4 Case study development

Case studies act as a mechanism for injecting reality into a course. They describe actual situations and pose real problems. As seen from the point of view of the decision maker they frequently identify a range of alternatives, a major point being that there is often not one answer. A key opportunity is therefore provided for students to exercise skills in decision making. A good *case study* for educational purposes is not a well defined problem with a specific solution. This contrasts with what might be presented as a *case history*: the latter could describe how to handle a situation and demonstrate best practice or, alternatively, how *not* to do things. We have concentrated on case studies rather than case histories.

Case study development has been largely driven by needs and the availability of materials. A major opportunity was presented by the need to develop the Department's flagship MSc course in Clean Technology. The involvement of local staff from Procter and Gamble in connection with product design initiated the development of the first case study related to sustainability. The second study arose from intense local interest and debate concerning the Byker waste reclamation plant in Newcastle. A third study was evolved from some research undertaken in collaboration with AMEC Process and Energy and Shell Expro on clean design issues, using as the basis data from the design of the Shearwater oil and gas platform which was constructed at AMEC's Tyne yard. Further studies have followed on from our contacts with the Environment Agency on the subject of Sustainable Transport.

While these developments led to enhancement of Masters level courses, including the undergraduate MEng, it has become clear that there is a need to develop simpler case studies for the earlier undergraduate years. This latter activity is now in hand, an example being the development of a study based on the disposal of sewage sludge which has been piloted with first year undergraduates. This is a fine example of inter-institution dissemination as the material was derived from the work done by Charles Duff, the VP at Surrey University.

The case study material and associated lectures currently being used are evaluated through cycles of use and feedback in the modules in which the material is deployed. It has become apparent that a range of   presentation media is essential if the material is to be used effectively. For example, a novelty was introduced in the Byker case study where small groups of students make their own video presentation of their viewpoint on the issue. Current work is to look at using technology such as the Web for dissemination, and for distance learning based on the case studies.

D002/009 © IMechE 2002

4 THE CASE STUDIES

4.1 Product design

The case study is presented in the form of a workshop and is an exercise in incorporating the principles of sustainability into product design decisions. The activity concerns the design of a domestic washing product. Case study material, provided by a leading manufacturer (Procter and Gamble), and information from the public domain have been combined into a workpack accompanied by tutor notes. Students receive an introductory lecture and a briefing on the workpack. They are formed into teams of five or six and given two weeks to work on the design. After one week a workshop is held to discuss ideas. After a further week, each team presents a group report and makes an oral presentation to the class.

The case study documents consist of:
1. The Task and Task Summary Sheet.
2. Background Notes:
3. Selection criteria for: Detergent Form (Powder, Tablet or Liquid), Main Surfactant, Main Builder, Additives.

The Task and Task Summary Sheet contains the instructions to the workshop students and a sheet for the students to record design decisions and notes on the reasoning used to justify those decisions. Background Notes give introductory and general information on aspects of household laundry detergent design. Other documents cover the options for design decisions. Students are encouraged to read all the documents provided, conduct their own research and meet as a group to discuss the task before the first workshop.

The first workshop session involves discussion within individual groups and selection by each group of its set of preferred options for the design of a household laundry detergent. The session is supervised by the facilitators. In the second workshop each group makes a presentation to the class (about 20 minutes long), outlining the group's design decisions and the justification of those decisions. This is followed by discussion of each presentation within the workshop class as a whole. Each group then submits a detailed report based on the workshop task, allowing an opportunity for groups to revise their design decisions as a result of workshop feedback.

4.2 The social dimension

A workpack has been assembled from data in the public domain concerning the design and operation of the Byker waste treatment plant. This focuses particularly on social and environmental issues. Following a briefing, teams are formed and asked to role-play. Roles are, for example: technology providers, the Local Authority, the media, concerned residents, environmental campaigners, etc. Each team spends two weeks under guidance developing their 'case'. The teams are then booked into the University's Audio Visual Centre to make a 10 minute video presentation of their case. The videos are replayed and discussed in a subsequent workshop. As with most of the case studies, there is an excellent opportunity with the Byker study for students to practise key communication skills, which enhances the value of the exercise. Each group is responsible for organising itself.

The final report from each group consists of a short outline of the group's position accompanied by the minutes of the group meetings (except 'Local Newspaper' who do a Press Release). The outline is accompanied by a critical assessment of the role a group played in the workshop, *viz.* application of methods and principles of Sustainable Development and an appraisal of how the group's case and proposed actions would lead to a more sustainable solution.

4.3 Decision making for Clean Design

This case study is based on data provided by an engineering contractor involved in the design and construction of a major offshore oil and gas facility. It is based around Environmental Design Objectives and considers a number of technical options for the treatment of 'produced water' from production wells. The workshop begins with video clips made by the contractor, followed by two lectures on decision making in theory and in practice. A workshop is then held in which 'design teams' are formed where individuals role-play (e.g. process engineer, environmental engineer, mechanical engineer, etc.). Because the module where this case study is used has students from a diversity of academic disciplines, it gives each participant an opportunity to exploit his or her particular expertise. Eight different designs are presented to each group for evaluation in terms of their sustainability – in effect the groups act as *design review teams*. Each team considers the design options in a structured way and applies the decision making methodologies. The options and their rankings are then discussed at a subsequent workshop. The exercise is supported by a workpack describing the background to the disposal of 'produced water' offshore, and a detailed description of each design option.

4.4 Systems specification

This case study was developed as a relatively uncomplicated introduction to a module on Design for a Sustainable Environment. It concerns the specification of refrigeration or air conditioning equipment for a selection of scenarios. Scenarios include: a design team working on a 'flagship' public building, a manufacturer of refrigeration systems, an operator of industrial scale refrigeration, facilities management of a supermarket chain, etc. Students are formed into small teams and given the task of specifying the conceptual design requirements for refrigeration systems for designated scenarios. The aim of this case study is to encourage the students to apply some creative thinking to find a solution to a practical problem. They are required to find a solution in terms of a conceptual system specification rather than detailed equipment design. The case study is based on information which is available in the public domain and students are required to do a comprehensive literature search using the world-wide web and conventional literature sources. The module is introduced by the facilitator who presents the background to refrigeration and refrigeration technology in a series of PowerPoint slides. The requirement is for students to provide a coherent and convincing case for a system design for a particular given scenario. They are judged on their application of sustainable design principles in finding a solution. The case study pack consists of the PowerPoint slides, tutor's notes, and an assignment task sheet.

D002/009 © IMechE 2002

4.5 Sustainable Transport

The workshop explores sustainable development in the context of cars and private passenger transport. The workshop entails group work, with each group representing a stakeholder with a distinct approach to the subject. The students are given background notes, supplementary documents, tables of data and other information which is available online to students or as printed notes. The tables give data about issues such as air emissions, travel statistics, etc. The roles and tasks are:

- *Vehicle Manufacturer*: develop an outline design for a sustainable, commercially viable family car that could be put into production within 4-5 years.
- *Central Government*: develop policy instruments, including fiscal measures if appropriate, with the aim of encouraging sustainability in car design and use.
- *Research Funding Body*: decide where to target funds to encourage the development of sustainable car design that will be feasible within the next 20 years.
- *Environmental NGO*: describe its vision of its ideal, sustainable form of private, passenger vehicle.
- *Company Car Fleet Manager*: choose a selection of car models that will make up the most sustainable fleet.

4.6 Sludge disposal

This case study was developed by drawing on material that had been developed by the VP based at Surrey University (Charles Duff) and relates to experiences with the Bran Sands sewage treatment facility on Teesside. It was presented as a simple introduction to sustainability for first year undergraduates in Chemical Engineering at Newcastle. The task is to consider the most sustainable final use or disposal method for sewage sludge.

The workshop which used the case study followed the usual pattern of presentation by the facilitators coupled with small group discussions. The student groups then had a week to develop their ideas and carry out research in preparation for an oral presentation the following week. Information was provided on sewage treatment and the nature of sludge. Options for the final use or disposal for sewage sludge were presented, together with the arguments for and against:

- Landfill
- Sea dumping
- Agricultural land application
- Sylviculture and forestry application
- Land reclamation
- Incineration with energy recovery
- Gasification

The aim was to introduce ideas of tradeoff, decision making and the 'triple bottom line'. It was also intended to make the students aware that there is rarely one correct (or even a viable) solution to such a problem and that part of the excitement of engineering is the need to use balanced judgement.

5 DELIVERABLES FROM THE PROJECT

The intention is to assemble the material as 'shrink-wrapped' packs including resource information, problem statement, worksheets and a facilitator's guide. In addition, during the supplementary two years of the programme at Newcastle a major part of the work will concentrate on providing web-based dissemination and material in a form suitable for the support of distance learning. It could also be argued that there have been some intangible benefits which are difficult to measure, such as the involvement of staff who might not have been otherwise engaged. To some extent, the programme is as much about raising awareness with staff as with the students.

6 OUTCOMES

The two original case studies (washing product, Byker incinerator) are now well-established and used routinely in the MEng/MSc Clean Technology module. The Byker case study has also been tested at Oxford and Aston universities with some success. Both of these institutions have modified the case study, and the way it is presented, to suit their own purposes. Whereas the Byker case study at Newcastle is being run with MSc/MEng level students, the workshop was modified for the Oxford trial to be suitable for first year Engineering undergraduates. Another change was that at Newcastle the workshop runs over three weeks (introduction, in-class workshop, report-back session) whereas the Oxford trial ran full time over 1½ days.

The evaluation of the programme is necessarily *ad-hoc* at this point in time. It is intended that a more formal evaluation will be carried out before the programme finishes in 2003. Nevertheless, we have pleased with the enthusiasm and seriousness with which the participating students have approached the case studies. A particular advantage of the case study approach is that apart from the inherent raising of awareness of sustainability issues, it has enabled students to practise their key skills of communication, presentation, decision making and teamwork.

The Academy initiative has helped us move sustainability firmly onto the curriculum. We have also been able to move away from a teaching (ie. lecture) based approach to a more student centred learning environment. There are, though, some negative points. Case study based learning has proven to be demanding on staff time and numbers. With the phasing out of the VP support in the near future, this situation will become worse. It is also dependent on the enthusiasm of staff. However, on the positive side, the material has been produced in a form which should be easily transferred to other student groups within the University and through the Royal Academy of Engineering to other HE establishments. It is our firm belief that, in future, students will increase their market value as a direct result of this raising of their awareness of the importance of the need to develop a more sustainable approach to problem solving where consideration of the 'triple bottom line' (economic, environmental and social aspects) becomes the key factor in all successful business decisions .

D002/009 © IMechE 2002

REFERENCES

1. **McQuaid J.,** (2000) ROAMEF Framework for a Scheme for Visiting Professors in Engineering Design for Sustainable Development, Royal Academy of Engineering

2. **Ali-Khan S.,** (1996) Environmental Responsibility Report Review (Toyne II) Forum for the Future

ACKNOWLEDGEMENTS

The authors would like to thank the following Research Associates who contributed significantly to the development of the case studies: Sharon Joyce, James Marshall, Lise Stoyell.

We are also indebted to the Royal Academy of Engineering who funded this work and who have provided administrative support to co-ordinate the efforts of all fifteen universities who are involved in the Visiting Professors scheme.

Appendix

Project Deliverables

Case Studies and Workshops addressing Sustainable Development issues

The following case studies have been developed. In addition a case study written by the University of Greenwich on Brent Spar was used in 2000 for a class of 55 Chemical Engineers. Is it very possible that other Departments and Faculties at Newcastle University as using a case study approach to the teaching of Sustainable Development. Information provided here is based on case studies used in Chemical Engineering.

Subject	Degree course (s) where used	Number of students involved
Design of a Sustainable Detergent, (developed in conjunction with Procter and Gamble)	MEng Chemical Engineering BEng Chemical Engineering (Erasmus) MSc Clean Technology MRes Responsive Processing MRes Marine Technology	50 per year, (run over previous three years, 99-02). Total 150
Disposal options for Newcastle City's Municipal Waste (using the Byker Incinerator case as background).	MEng Chemical Engineering MSc Clean Technology MRes Responsive Processing MRes Marine Technology	50 per year, (run over previous two years, 00-02). Total 100
Sustainable motor transport design	MRes Marine Technology MSc Clean Technology BEng Materials, Manufacturing and Mechanical Engineering MEng Materials, Manufacturing and Mechanical Engineering	65 per year (ran first time in 01/02)

Design of a Sustainable Refrigeration System	MEng Chemical Engineering BEng Chemical Engineering (Erasmus) MSc Clean Technology MRes Responsive Processing	50 per year (ran first time in 01/02)
Consideration of sludge disposal options	BEng Chemical Engineering (Year 1)	50 per year (ran Feb 2002)
Assessment of local social economic and environmental impact of climate change	BEng Chemical Engineering	50 per year (ran Feb 2002)
Disposal options for process water from an oil & gas platform	MEng Chemical Engineering BEng Chemical Engineering (Erasmus) MSc Clean Technology MRes Responsive Processing	50 per year (ran Feb 2002)
Design of a sustainable water treatment system	MEng/BEng/MSc Design project for Environmental Engineering and Clean Technology	60 per year
Case study on foot and mouth animal disposal	MEng, BEng Environmental Engineering, MSc Clean Technology	60 per year (to run spring 2002) – developed by the Civil Engineering Department

Educational challenges of web-based case studies in sustainable development

M HUTCHINGS, S LEWAREN, and **K NORMAN**
Academic Services, Bournemouth University, UK
N GARLAND and **M HADFIELD**
School of Design, Engineering, and Computing, Bournemouth University, UK
G HOWARD
Smith & Nephew plc, Birmingham. UK

ABSTRACT

The Royal Academy of Engineering has funded a project to produce case study teaching material relating sustainable development issues to undergraduate students within the School of Design, Engineering and Computing at Bournemouth University. This is a challenge because it is a complex wide-ranging subject matter and there is no unique solution. The principal learning vehicle is an interactive case study approach available to students through the Internet. This paper addresses the specific educational challenges of student-centred web-based learning and teaching strategies applied to the complex subject of sustainable development. Pedagogical components of variety, action, application, interaction, assessment, feedback and evaluation are discussed and related to the learning and teaching of the subject using the web. Life Cycle Analysis (LCA) software is included as a teaching tool and its role when interfaced with case studies is discussed.

1 INTRODUCTION

This paper describes the educational challenges of learning and teaching sustainable development to undergraduate design students at Bournemouth University [1]. An interactive web-based learning environment is adopted to introduce concepts and quantify environmental impacts using Life Cycle Analysis (LCA). The student designer has significant control over material selection and the end of use strategy and hence these are highlighted in the learning philosophy. An expert in sustainable design for waste minimisation from Smith & Nephew plc is funded by the Royal Academy of Engineering as a Visiting Professor to develop this learning opportunity with design education and web-based learning specialists at Bournemouth University.

1.1 Sustainable development rationale in design education
Awareness, understanding and integration of sustainable development issues within the design process are the underlying focus. A holistic philosophy is the general ethos of design courses at Bournemouth University and thus sustainable development is a key issue. Specifying products that meet the needs of the present generation, without compromising the ability of future generations to meet their own needs [2] is the challenge for designers. Sustainable development needs to address issues that include economic growth and employment, social progress for all, protection of the environment and prudent use of natural resources. This is a complex and multidisciplinary subject and thus modern information technology is embraced to facilitate and communicate effective industrially relevant case studies.

The project aims to introduce, reinforce, and develop understanding of sustainable development concepts based on a multidisciplinary approach. A greater challenge is to embrace a web-based case study approach to enable design students to implement sustainable development criteria within the design process. The intended learning outcomes are that after using the interactive case studies and supporting modules, the students will be able to demonstrate:

1) an understanding of the concepts of sustainable development and its framework and principal facets involving eco 'green', legislation and technical issues;
2) an understanding of the design process context within the sustainable framework (commercial and institutional drivers for sustainable development);
3) knowledge of practical problems and solutions within industry to achieve sustainable products and processes;
4) the means to implement and influence sustainable development within the product/engineering design process.

To facilitate these learning outcomes, a web-based learning environment has been developed [3] concurrently with conventional teaching methodology ie lectures, seminars and design project awards (Smith and Nephew Prize for Sustainable Development).

1.2 Interactive Case Study Rationale
An important issue is implementing sustainable design thinking and quantitative tools effectively within the product design process. Design students should also be aware of the many environmental impacts and be able to decide on primaries for a particular design. Design for disposal, the link between technology, ecology and economics should be appreciated (all can be winners) and a range of products, processes and materials considered. It is evident that modern information technology needs to be embraced to facilitate and communicate effective industrially relevant case studies. It is clear that principles from expert knowledge based systems such as commercial life cycle analysis techniques should be encouraged.

2 EDUCATIONAL CHALLENGES

So what does this mean for the teacher and the teaching strategies adopted to create an interactive web-based case study? Teaching is about changing the ways in which learners understand, experience or conceptualise the world around them as a result of a variety of learning experiences [4]. Teaching methods adopted within the web-based case studies will have different impacts on learning outcomes just as they might within a conventionally designed learning environment. Teachers' conceptions and approaches to teaching have been found to be closely related to those describing conceptions of learning [5]. Prosser and Trigwell's model [5] identifies some teachers as holding a teacher-centred view, focusing on the content, syllabus, and textbooks, and see their role as transmitting information based on their knowledge. At the other extreme, teachers focus on their students in planning their activities and see their role as helping their students develop and change their conceptions. These teachers are likely to adopt different approaches to teaching with consequent effects on the quality of the student learning experience. The teacher's conceptions have a powerful influence on what and how students learn, with students adopting deep or surface approaches according to the demands of the task [6]. Our underlying educational philosophy is summed

D002/011 © IMechE 2002

up in Laurillard's statement that: "It is the teacher's responsibility to create the conditions in which understanding is possible, and the student's responsibility is to take advantage of that" [7].

Our web-based case study approach sees the role of the teacher as interpreting the wide-ranging subject matter of sustainable development and providing scaffolding to support and enhance the student learning process. The educational challenge for the design team was to consider what would enable the students to learn rather more than what to teach them. Rowntree argues: "It may be better for learners to fully understand half the content you would ideally like to include than to half understand all of it" [8 p.59]. Ramsden recommends teachers "to reflect on what we do that influences the relation between the students and what they learn" [9 p.158].

The Visiting Professor provided the subject content and learning resources and had to wrestle with what to include and what to exclude in order to assist students in interpreting the information and learning by doing. The design team had considerable choice in the range of pedagogical components that could be employed within the learning design of the web-based case studies. The educational philosophy underlying the team's approach was to design a learning package which put the student at the heart of the learning process. Our purpose was to actively engage the students in critical thinking and problem solving rather than simply reading pages of information in electronic form. This is achieved by using student-centred pedagogical design components - the specification of learning outcomes for the students, guided activities to encourage students to learn by doing, and self-assessment questions - interlaced with subject information.

Learning outcomes are described in a positive, shared and informal style (Figure 1). Advance organisers are used to alert students to what is to follow and summaries are included to help the students to digest the main points (Figure 2).

Introduction to your Life Cycle Analysis Task

What will you learn?

- Understand how the environmental impacts/burdens can change in selecting paper, plastic or metal materials for your design
- How transportation can influence the impacts
- How recycling can change the situation
- Encourage you to design for re-use, recycling
- To use the minimum amount of correct material

Figure 1. Learning outcomes for LCA Task [3]

Life Cycle Analysis – Case Study Progress

So far we have looked at the concepts of Life Cycle Analysis and identified the value of considering LCA within the design process of several mass-produced products (motor car, washing machine and toothpaste) and the importance of using materials appropriately.

Figure 2. Summary for Case Study Progress [3]

Student-centred learning activities form the core of the case study with self-assessment quizzes (Figure 3) and a number of self-reflective questions embedded in the text (Figure 4). These build and guide students towards the Life Cycle Analysis task using Eco Packager, PIRA's Streamlined Simple Life Cycle Assessment Model [10].

Life Cycle Analysis Quiz

Objective – To check your understanding of the first part of the case study and the concepts of Life Cycle Analysis

Question 1
Rank the following major environmental burdens that should be considered in the Life Cycle Analysis of a product or process. Start with the most important burden in position 1 through to the least important burden in position 7.

Options
- Global warming
- Ozone depletion
- Human toxicity
- Acidification
- Water quality
- Damage to structures
- Ecotoxicity

Figure 3. Life Cycle Analysis Quiz: Sample Question [3]

Goal and Scope Definition of an LCA

Practice example

For example, should the saw used to cut the trees be included in an LCA of paper production? If so, the environmental burdens of its use need to be identified. This would require an assessment of the components used in its manufacture and operation, the fuel for its operation, etc. An appropriate cut-off point or boundary has to be defined.

Figure 4. Reflective Question embedded in text [3]

What does this pedagogical approach mean for the student engaging with these web-based case studies? Some general points concerning the educational value of web-based learning are applicable here. Web-based learning, like computer assisted learning, enables students to work at their own pace. It is available at the students' convenience – any time, any place - provided there is access to a networked computer. It ensures consistency of content and presentation. It has interactive potential and it can have novelty value. Interactive potential is a key consideration for designers of web-based learning. Forsyth argues we could "...end up with electronic page-turning, limited or no interaction and the inability for the teacher and learner to interact in real-time or even communicate" [11 p.31]. The novelty value depends on the impact and quality of the page design and the user interface as much as on the pedagogical components included in the learning design. Used well, it can help to generate and maintain student interest and motivation.

D002/011 © IMechE 2002

The intended outcome of this web-based case study approach is for students to learn but as with conceptions of teaching, conceptions and approaches to learning will inform the learning design and the possible learning outcomes. Students go about their learning by using different strategies, with corresponding learning outcomes and levels of success. Early work by Perry [12] indicated a developmental trend in student thinking from a belief in *dualism*, where right answers exist, provided by teachers, for producing in assessments, to an awareness of *relativism*, where conclusions are based on evidence to be cautiously interpreted by learners. Research by Säljö [13], Marton [14] and Beaty [15] led to the identification of six categories of learning. The categories form a hierarchy with higher conceptions based on the foundations of lower conceptions. The lower conceptions see learning as something external to the learner, a product, which can be taken and absorbed by the learner. The learner demonstrates achievement of learning outcomes by being able to *reproduce* the information. By contrast, the higher conceptions identify the role of the learner in making sense of the meaning, in internalising the learning and understanding it in the context of the real world. Here the learner demonstrates achievement of the learning outcomes by being able to *transform* the information.

Individual conceptions of learning affect approaches to learning. Marton and Säljö's work [16] [17] generated two key definitions based on students' descriptions of their approaches to learning, a *surface* approach and a *deep* approach developed by Gibbs [6]. Positive learning outcomes can be demonstrated where it can be shown that the student is able to transform the information. The intention of the web-based case study approach is to both support and challenge the student. This is encouraged by providing them with active learning opportunities through questions and tasks supported by information and feedback. The aim is to encourage students to achieve deeper learning, which can be demonstrated in the achievement of higher level learning outcomes.

Key learning design principles were used to assist the design team in their selection of pedagogical components and related page design devices intended to enhance the student learning experience. These included the requirement for:

Variety: Learner's attention engaged with range of stimuli
Action: Learner is doing things
Application: Learner encouraged to apply learning in another context
Interaction: Learner can change/comment on content
Feedback: Learner reflects on what they have done or understood
Evaluation: Learner contributes to learning design through feedback on the learning experience.

The intention is to analyse each of these key design principles in turn by explaining the range of pedagogical components and page design devices adopted to fulfil these principles.

2.1 Variety: Learner's attention engaged with range of stimuli
The text is presented in short paragraphs with use of sub-headings as navigational cues. The amount of text per page is limited to ensure scrolling is kept to a minimum. The text is interspersed with photographic illustrations, schematic/block diagrams and thumbnail sketches capable of enlargement. These devices help to chunk the material into manageable form.

Links to other resources from within the text are kept to a minimum to avoid students wandering off to other sites. Hypertext links are provided within a section at the end of the case study entitled Sources of Further Information.

2.2 Action: Learner is doing things
Questions and prompts are embedded within the text to encourage the student to engage actively with the content. Students are presented with opportunities to review their learning by means of self-assessment quizzes, the first reviewing their understanding of the case study and the concepts of LCA and the second encouraging them to reflect on environmental burdens as a part of the design process. Students are asked to make choices from a range of tick box and pull-down menu options.

Navigation is provided by forward and backward links through the pages but learner choice is guided along a linear path through the pages. An index is provided at the end of the case study to allow learner choice in selecting areas for review and revision. Mouse clicks are required to move from page to page.

2.3 Application: Learner encouraged to apply learning in another context
The student is provided with illustrated examples within the study. These examples enable the student to identify and compare notable environmental impacts of a number of products, for example, motor cars, washing machines, and toothpaste packaging. The self-assessment quizzes allow students to apply their learning in answering the questions set.

Having successfully completed the case study, the student is asked to undertake the key Life Cycle Analysis activity using the Eco Packager program. The student is tasked to design a simple product, a tomato sauce bottle. They are asked to make decisions about the materials to be used in their designs and to calculate individual weights of raw materials for their product. They are asked to consider transport for their product and disposal options at the end of its life.

2.4 Interaction: Learner can change/comment on content
The Eco Packager program enables students to model different materials selection and transportation and disposal options. The student is asked to go through two iterations of the design process making changes to their selection of materials, process routes, waste strategy, and re-cycle strategy to enable them to analyse the effects of their decisions on environmental impacts. The data based on their design decisions is input to the program and they receive immediate feedback with results presented in tabular form showing the environmental impacts of their choices. Students are encouraged to prioritise environmental impacts and show different design outcomes.

Links to the Glossary and Sources of Further Information enable the students to make connections between different aspects of their learning. It is possible to link to these at any stage by means of the navigation bar standard at the top of each page.

2.5 Feedback: Learner reflects on what they have done or understood
The self-assessment quizzes provide important opportunities for the students to review their learning. Instant feedback is given and student progression is dependent on reaching an agreed level of competence in the quizzes. Although feedback is based on three score bands from successful, through competent, to unsuccessful, students do not see their scores. They

receive encouraging and supportive feedback statements with associated links, directing them along appropriate learning pathways according to their feedback scores, forward if they scored well, back to review earlier information if their score was low, and choice in either direction if they reached the level of competence. Where students continue to experience difficulties after reviewing the materials and repeating the quizzes, they are recommended to seek help from their tutor.

Learning outcomes are demonstrated by the learner being able to transform the information with which they are engaging. The product of the student's learning is submitted to the tutor for formal assessment. In this way the activity of each case study is formally presented by the student and monitored by the tutor.

Additional feedback devices are planned including a direct e-mail link to the tutor for questions, a frequently asked questions database available within each case study, and an online discussion forum.

2.6 Evaluation: Learner contributes to learning design through feedback on the learning experience

An online questionnaire included at the end of the package provides the opportunity for students to give feedback on their experience of using the package. It includes eleven statements, which students are asked to rate on a five point scale from "strongly agree" to "strongly disagree", together with opportunities for qualitative comments. Student evaluation of the case study has been very positive with comments like "The exercise was well worked and informative". All students responded with "agree" or "strongly agree" to the statements "The aims and objectives of this web-based package are made clear" and "I am now able to identify the purpose and function of Life Cycle Analysis". Students were positive about the statement: "The quizzes helped me review my learning", with only one student "neither agreeing or disagreeing". Students were also positive about the ease with which they were able to complete the assessment task with only one student disagreeing. In response to the question: "I need more help in:....." one response asked for "More explanation to how to complete the task" and in response to the question: "I would like to see more information on....." one response said "How to understand the charts". However all but one student "agreed" with the statement "I can now justify the role of Life Cycle Analysis in product design".

3 IMPLEMENTATION ISSUES IN USING THE WEB FOR DELIVERY

The widespread adoption of the Internet within British society has given rise to a series of issues for online learning. In a positive vein, the Internet has done much to improve access to a wealth of high quality information. The proliferation of irrelevant information and the expense of time online are two factors that have adversely shaped the patterns of use by users.

Online users make rapid judgements of relevance and interest. Additionally, the ease of switching to an alternative source of information has given rise to a demanding user group. Sustainable development is a very complex subject with the need at some stage to make a balanced decision based on a great deal of information gathered on the environmental, social and economic aspects of the product/engineering design. The issues of presenting the

Sustainable Product Development message in an interesting, learner-centred format were therefore clear from the outset.

Initial consideration of the requirements for the website suggested that a consistent look and feel would be needed throughout. The development of a suitable logo set the tone for much of the development work. Organic colours such as green and brown were employed extensively to reinforce the ecological sensibilities behind the sustainability concept. The same colours were also combined in the finished logo to suggest an 'end of life' decay in the form of rust. Wherever possible, the range of colours used within the website was taken from a "web-safe" colour palette. In theory these colours should be displayed consistently across different user platforms regardless of the equipment and software used to view the website.

The use a graphical navigation banner from the beginning of the case study offers the student control over their progression through the information provided. It is possible to move forwards but also backwards to review material already seen. Links to call up sources of further information and a glossary can be selected at any time to provide explanations and access to manufacturer's websites.

The need to create impact was the main consideration behind the welcome page. It was felt that it should provoke a realisation within the student's mind of the importance and far-reaching consequences of the subject. A series of striking images were selected to convey the relevance and pervasive nature of the sustainability issue (Figure 5).

Figure 5. Sustainable Product Development Welcome Page

From the outset of the project it was hoped that a series of interactive and graphical questions could be incorporated to provide interest and variety. Unfortunately this proved a greater challenge than had been anticipated due to software problems. A decision was made to withdraw the problematic questions and replace them with text only versions of the same.

The final challenge was to integrate an existing modelling program [10] within the case study. Linking to the program was relatively straightforward but providing sufficient instruction prior to use provided a greater test. It was not feasible to modify the program so a decision was made to provide printable pages to assist users and allow them to record the results of their investigation for the assessed task.

Despite limitations in the performance of software and issues such as cross-platform compatibility, the majority of the original ideas for the website were implemented. With such projects it is often a case of considering the typical software and hardware configurations available to university wide audiences.

4 CLOSING REMARKS

This paper has attempted to demonstrate the educational challenges surrounding the content and learning design of a student centred web-based learning and teaching approach. The subject of sustainable development is a complex and wide-ranging one. Choosing methods and techniques to encourage deep learning on the part of the student is as challenging as decisions about what information to include.

The experience of designing web-based case studies has proved a challenge for all of the team. Team members have contributed their different areas of expertise to the content, learning design and implementation of this web-based learning environment. Much has been learnt from each other in the course of the project design and implementation. It is hoped that this paper will serve to provide a guide for others embarking on web-based learning developments. The educational challenges of providing an interesting and motivating learning experience, combining support and challenge for the students, is not an easy one, but one that can be embraced with the close co-operation of a team of subject and web-based learning specialists.

REFERENCES

1. **Royal Academy of Engineering** (1998) Visiting Professors in Engineering Design for Sustainable Development
   http://www.raeng.org.uk/education/vps/sustdev.htm

2. **The Brundtland Commission** (1987)

3. **Bournemouth University** (2002) Sustainable Product Development Website
   http://spd.bournemouth.ac.uk

4. **Ramsden, P.** (1992) *Learning to teach in higher education*. London: Routledge.

5. **Prosser, M.** and **Trigwell, K.** (1999) *Understanding learning and teaching: the experience in higher education.* Buckingham: Society for Research into Higher Education and Open University Press.

6. **Gibbs, G.** (1992) *Improving the quality of student learning.* Bristol: Technical and Education Services.

7. **Laurillard, D.** (1993) *Rethinking university teaching: a framework for the effective use of educational technologies.* London: Routledge.

8. **Rowntree, D.** (1990*) Teaching through self-instruction: how to develop open learning materials.* London: Kogan Page.

9. **Ramsden, P.** (1989) Perceptions of courses and Approaches to Studying: an encounter between two paradigms. *Studies in Higher Education* 15 (2) 155-168.

10. **PIRA International** (2001) Eco Packager
http://www.lcauser.com

11. **Forsyth, I.** (1998) Teaching and learning materials and the Internet. 3rd ed. London: Kogan Page.

12. **Perry, W.G.** (1970) *Forms of intellectual and ethical development in the college years: a scheme.* New York: Holt, Rinehart and Winston.

13. **Säljö, R.** (1979) Learning in the learner's perspective. I. Some common sense conceptions. *Reports from the Institute of Education,* 76. Gothenburg: University of Gothenburg.

14. **Marton, F. et al.** (1993) Conceptions of learning. *International Journal of Educational Research* 19 277-300.

15. **Beaty, E., Dall'Alba, G.** and **Marton, F.** (1997) The personal experience of learning in higher education. In Sutherland, P. (ed.) *Adult Learning.* London: Kogan Page, pp. 150-152.

16. **Marton, F.** and **Säljö, R.** (1976) On qualitative differences in learning - I: outcome and process. *British Journal of Educational Psychology.* 46, pp. 4-11.

17. **Marton, F.** and **Säljö, R.** (1997) Approaches to learning. In Marton, F., Hounsell, D.J. & Entwistle, N. (eds*.) The experience of learning.* Edinburgh: Scottish Academic Press.

D002/011 © IMechE 2002

Authors' Index